the CBD beauty book

the CBD beauty book

Make your own natural beauty products with
the goodness extracted from hemp

Colleen Quinn CA

CICO BOOKS
LONDON NEW YORK

For Katie and Ava—you hold your dreams in the palm of your hands, little darlings!

Published in 2021 by CICO Books
An imprint of Ryland Peters & Small Ltd
20–21 Jockey's Fields   341 E 116th St
London WC1R 4BW   New York, NY 10029

www.rylandpeters.com

10 9 8 7 6 5 4 3 2 1

A CIP catalog record for this book is available from the Library
of Congress and the British Library.

ISBN: 978-1-80065-020-6

Printed in China

Designer: Geoff Borin
Photographer: Joanne Murphy
Commissioning editor: Kristine Pidkameny
Senior editor: Carmel Edmonds
Art director: Sally Powell
Head of production: Patricia Harrington
Publishing manager: Penny Craig
Publisher: Cindy Richards

Important safety notes

• Readers are urged to consult a relevant and qualified specialist
or physician for individual advice before taking CBD in conjunction
with any other medical conditions or medication, or if pregnant
or breastfeeding.

• Please note that while the descriptions of essential oils and
blends refer to healing benefits, they are not intended to replace
diagnosis of illness or ailments, or healing or medicine. Always
consult your doctor or other health professional in the case
of illness. Neither the author nor the publisher can be held
responsible for any claim arising out of the general information
and blends provided in this book.

Contents

Continued overleaf

Foreword by Ethan Russo, MD

It is a pleasure to pen this foreword to Colleen Quinn's lovely book on cannabidiol (CBD) and skin care.

Cannabidiol is the second most abundant cannabinoid in commercial cannabis and number one in industrial hemp. For decades after its isolation by Raphael Mechoulam in 1963, CBD was largely unnoticed in comparison to tetrahydrocannabinol (THC), its (in)famous psychoactive counterpart. Researchers in Brazil kept CBD's torch alive by investigating its potential to treat epilepsy, anxiety, and other disorders in humans.

The thread was picked up again in earnest in the late 1990s, when GW Pharmaceuticals actively pursued a breeding and research program on CBD and elucidated its myriad benefits in allaying inflammation, and its beneficial effects on pain, seizures, and countless other conditions. Along the way, science confirmed what was previously claimed by Renaissance herbalists: cannabis-based medicines containing CBD were effective in treating burns, scars, and arthritic conditions when applied to the skin. Subsequently, the antibiotic benefits of CBD and its ability to combat acne by reducing sebum production were documented. Its antioxidant effects were demonstrated in the brain and certainly apply equally to the skin. While most medicines are developed with one target in mind, CBD is a standout in having 30 or more mechanisms of action, making it a much sought-after ingredient in an increasing number of medicines throughout the world—all this combined with a uniquely benign side-effect profile.

CBD combined with Colleen Quinn's beauty (and therapeutic!) recipes make for a wonderful marriage. Her formulations are sumptuous in their allure and healing promise. This is the book on skin care that I have always wanted, combining all-natural ingredients into blends that will nourish the senses, calm the soul, and heal the face and body we present to the world. Experiment and savor the results!

Ethan Russo, MD
Neurologist, cannabis researcher, and CEO of CReDO Science

Introduction

My mission is to share the power of essential oils with as many receptive minds as possible. These remarkable natural products have been my life's work. For the past 20 years, I have studied and applied their chemistry, bioactivity, properties, and clinical and cosmetic uses.

In 2015, cannabis entered my life like a bolt of lightning! Cannabis came with an enormous difference from any other plant I had worked with up to that point. It has a completely unique affinity with the human brain. Since 1988, it has been known that the brains of mammals have receptor sites that respond to compounds found in cannabis called cannabinoids, one of which is CBD. Archaeological evidence for the medicinal use of cannabis before recorded history began tells us that our distant ancestors intuited the ability of the cannabis plant to activate the natural cannabinoid responses of the body. However, it was not until 1992 that the ongoing study of cannabinoids led to the groundbreaking discovery of the human endocannabinoid system (ECS). I touch on the implications of that discovery a little later (see page 15).

I became enthralled by cannabinoids, especially CBD. Their synergistic interaction with the ECS elevated my work to a new and unforeseen level. I designed and taught courses on the therapeutic effects of CBD oil, and incorporated CBD when formulating skincare products for global brands. Fascinated by the results of ongoing research, I went to California in 2016 to learn more, working with enthusiastic and dedicated cannabis farmers in cultivating unique hybrids, while also formulating terpene-rich blends (terpenes are other compounds, found not only in cannabis but also in essential oils— see page 17) with different formats of cannabis for both the skincare and pharmaceutical worlds.

Nature's bounty is the greatest gift we have at our fingertips. Sometimes we just need a little guidance on how to use these highly versatile plants to treat, heal, and support our skin, body, and mind. I wrote this book for that reason.

For centuries, aromatherapists, herbalists, and naturopaths have been drawing on a boundless natural skincare catalogue of plants, clays, herbs, grains, and other nutrients. With today's prevailing movement toward green beauty and veganism, natural beauty support is even more popular and now, thankfully, more accessible to us all. Whether you are a passionate natural beauty advocate or simply curious about CBD and natural beauty, this book will give you a foundational understanding of what CBD is, and how it affects our skin and our body. It will take you on a voyage through some of the most powerful aromatic substances available to us from nature.

We begin with an introduction to CBD—what it is and how it works—and I will also advise on what to look out for when buying CBD oil for the skincare recipes in this book. As well as an explanation of how beauty products work with the skin, a guide to the botanical ingredients used in the 35 beauty recipes follows. Clear, comprehensive charts will equip you with an understanding of the therapeutic and cosmetic properties of all these ingredients.

You will then be ready to explore the wonderful collection of meticulously curated facial and body skincare recipes. All these beautiful recipes are infused with CBD oil and provide a powerful antioxidant and anti-inflammatory treatment for your skin. They are also vegan, nut-free, and designed to be kind to all skin types.

I hope you enjoy discovering the world of natural CBD beauty.

CHAPTER 1

Meet CBD: Your New Beauty Ingredient

You may be surprised to find cannabis in the pages of a book on natural beauty! The chances are that when you think of cannabis, you picture someone smoking a bong or a joint, on the way to feeling mellow. What you are less likely to know is that the cannabis plant was grown for medical use as far back as 3,000 years ago, and over the past few years has made a resurgence as a powerful plant-based medicine—to the extent that it has now been legalized for medical use in many countries across the world.

CBD is short for cannabidiol. It is the non-psychoactive component of cannabis and hemp plants. It has a truly daunting arsenal of therapeutic properties, being an antioxidant, anti-inflammatory, anticonvulsant, antipsychotic, anti-anxiety, analgesic, antidepressant, and neuroprotective natural plant compound. Many of these properties happen to be immensely helpful for skincare conditions, including aging, dehydration, dullness, blotchiness, acne, psoriasis, eczema, and excessive pigmentation. It is no wonder that this 100% natural wonder is achieving superstar status in the beauty and skincare world—especially when you combine it with other bewitching botanical ingredients, as you will discover in this book!

What is CBD?

First, let's look a little closer at this extraordinary compound. CBD is one of the most powerful parts of the cannabis and hemp plant. Unlike THC (short for tetrahydrocannabinol), which is the psychoactive chemical element of the plant, you can benefit from the therapeutic effects of CBD without any psychedelic effects.

CBD is what is called a cannabinoid. Cannabinoids are chemical compounds found in the cannabis plant. They closely resemble compounds called endocannabinoids that our own bodies produce naturally. It is considered to be one of the most effective compounds to help support our immune system and decrease anxiety and depression, as well as being an influential neuroprotectant, defending and supporting the vital neurons which make up our brain function. Two major contributors to the breakdown of brain cells are oxidation and inflammation. CBD has very robust antioxidant and anti-inflammatory properties.

WHAT IS THC?

THC can cause psychoactive effects, depending on dosage and previous exposure to the cannabinoid. However, it is also an impressive therapeutic compound which is being effectively used to treat a wide range of medical conditions and symptoms including pain, nausea, muscle spasms, appetite stimulation, anxiety, depression, and post-traumatic stress disorder (PTSD). Specifically, THC helps nauseous and sick patients to regain their appetite. Even patients who suffer from debilitating pain have seen effective results when medicating with THC. This is very encouraging, given the potential for addiction when treating with other forms of analgesia, such as opioids. The medicinal applications of THC are undeniable and, as a result, pharmaceutical companies are now creating synthetic versions of it to treat patients who suffer from the aforementioned conditions. However, as I said earlier, the focus of this book is firmly on the non-psychoactive CBD oil.

Given that so many recipes in this book combine CBD and essential oils, it is interesting to note that distinguished neurologist and pioneer in cannabis science Dr Ethan Russo has said that "research implies that the combined use of essential oils and cannabinoids may be a potential novel therapy for the treatment of neurodegeneration, and associated symptoms."[1]

One of the most widely accepted uses of CBD therapy is to treat patients who suffer from debilitating, chronic seizures. Have you heard about Charlotte's Web? It is a strain of medicinal CBD oil named for a young Colorado girl called Charlotte Figi. She suffered from a catastrophic form of epilepsy called Dravet syndrome. When she began taking CBD oil, she experienced a dramatic reduction in the number of her seizures—from around 300 a week to a handful a month. Charlotte became known as the girl who changed the medicinal cannabis laws across America. Sadly, she died in 2020, aged 13, from complications possibly related to coronavirus.

CBD covers a lot of ground while on its therapeutic travels. All types of inflammation, including digestive disorders and arthritis, may be improved through CBD-based therapy. However, one thing CBD will not do is get you high. It can relax you and ease anxiety, but it will not produce the impaired psychotropic effect that the cannabinoid THC provides.

Are you intrigued about these powerful plant compounds yet? Wait until you meet the endocannabinoid system! The science behind CBD is most interesting, but once you appreciate how it works with our body, CBD becomes positively enthralling.

How CBD Works

As CBD is absorbed by our skin, it makes its way to our endocannabinoid system and stimulates it, activating a series of processes beneficial to our health.

Above: **CBD oil can be added to beauty recipes for added benefits, such as in the Black Cumin Scalp Serum (see page 122).**

Meet the endocannabinoid system

The endocannabinoid system (abbreviated as ECS) is a remarkable network of compounds and receptors in the brain often described as a central component of the health and healing of every human and almost every animal. This vast grid has the capacity to influence functions in the brain, including memory, mood, pain response, appetite, perception, cognition, sleep, emotions, motor function, and anti-inflammatory function, as well as brain development and protection. The ECS is omnipresent in the body—in the skin, the brain, major organs, connective tissue, glands, immune cells, etc. In each area of the body, it carries out different tasks, but the goal is always the same and it is a rather wonderful one: the ECS works tirelessly to maintain the body's internal balance and physical wellbeing. It creates an internal equilibrium, harmony, and peace, which resists even the most hostile fluctuations in the external environment. This state is known as homeostasis (from two Greek words which mean "standing still").

Are you wondering why you have never heard of the endocannabinoid system before? If so, you are far from alone! The science is relatively new. In 1988, in a government-funded study at the St. Louis University School of Medicine, Allyn Howlett and William Devane determined that the brains of mammals have receptor sites that respond to compounds found in cannabis. These receptors, named cannabinoid receptors, turned out to be the most abundant type of neurotransmitter receptor in the brain. It was not until 1992, however, that the endocannabinoid system was discovered by Dr Raphael Mechoulam of the Hebrew University of Jerusalem, who was researching the cannabis plant at the time. It is almost unbelievable to think this discovery is so recent, considering that the ECS is so fundamentally important. As Dr Mechoulam says,

"There is almost no physiological system that has been looked into where the endocannabinoid system does not play a certain role."[2]

There are two types of cannabinoid: phytocannabinoids or plant cannabinoids (*phyto* is the Greek word for "plant") and endocannabinoids (cannabinoids that are produced naturally in the body—*endon* is the Greek word for "internal"). These two types are so incredibly similar that our body responds to them as though they are one and the same. At times when our body, on its own, does not produce enough endocannabinoids to maintain that desirable state of homeostasis, it will happily use phytocannabinoids to make up the deficit.

It is safe to say the endocannabinoid system is the controlling system of essentially all functioning within our body and mind. As the Italian researcher Vincenzo Di Marzo says, the ECS is "essential to life's basic processes by relaying messages that affect how we relax, eat, sleep, forget, and protect."[3]

The Cannabis Plant

So, we now know CBD is a cannabinoid and we have met the endocannabinoid system, but what about the plant? What is the difference between hemp, cannabis, weed, pot, grass, dope, ganja, and marijuana? Don't let the terminology spook you. Let me explain!

Every plant has a botanical name as well as a common name. Cannabis has two main species, *Cannabis sativa* and *Cannabis indica*. While they are both versions of the same plant, their properties are quite different. *Cannabis sativa* produces more CBD and less than 1% THC, whereas *Cannabis indica* is the drug-rich cannabis whose THC levels can be between 1 and 30%. *Cannabis indica* is the species which will get you high!

Cannabis and hemp

Cannabis sativa is the botanical name for both cannabis and hemp plants. The difference in how they are defined legally is the chemistry which makes up the DNA of each plant. Cannabis has a full array of cannabinoids and can contain anything from 1 to 30% THC concentration, whereas, although hemp contains all the cannabinoids, its THC level is always 0.3% or below. Pot, weed, marijuana, and other quirky names for the plant are all slang terms (along with many other professionals, I avoid using the word "marijuana" because of its historically racist undertones—see page 19).

Hemp is one of the oldest domesticated crops in the world. Throughout history, we have grown different varieties of this plant for various purposes, including to make clothing, rope, sails, biofuel, fibrous building material, plastic, and food. Today, some hemp plants are grown predominantly for their fiber, while others are grown primarily for their seeds to produce a carrier oil which is commonly used in beauty care and cooking (see right).

Cannabis plants are widely used for medicinal purposes. Unless cannabis is legal where you live, you will need a medical card, medical recommendation, or access to a legally approved dispensary in order to obtain medical cannabis.

Legally, hemp is the easier plant to access and use, because the THC levels are low and CBD levels tend to be high. It is CBD from hemp that is most commonly used in the beauty and cosmetic industries. You can buy CBD oil extracted from hemp online, in beauty or health stores, and even in supermarkets.

In the UK, CBD oil is extracted from hemp, whereas in Canada and the USA CBD oil can be extracted from either hemp or cannabis. You need to read the label to understand which variety of the plant your CBD oil has come from, but the CBD will be the same singular compound and have the same properties regardless.

HEMPSEED OIL

Many people buy hempseed oil under the illusion that it is CBD oil--but it is not. Hempseed oil is produced by pressing hempseeds, which do NOT contain phytocannabinoids. Let me be super-clear here: hempseed oil does not have any CBD in it—not even a token dose!

However, hempseed oil is a wonderful carrier oil for skincare recipes. It is stunningly therapeutic and rich in vitamin E (see page 27), and it holds a treasure trove of essential fatty acids (see page 29) which make it a restorative, regenerative, and replenishing anti-inflammatory and antioxidant skincare ingredient. It has the ability to protect and repair the skin from cellular damage while soothing irritation and balancing sebum production (see page 26).

Cannabis botany

Cannabis plants vary in height from 3 to 15 feet (1 to 4.5 meters) and have numerous branches with five to seven delicate jagged leaves that are spread like fingers. The plant in its entirety is covered with tiny, sticky, hair-like structures. These hairs are microscopic gland-like spikes that develop on the plant's skin. The technical term for these plant hairs is trichomes.

Trichomes are living cells. They protect the leaves and flowers and reduce evaporation by guarding the plant against wind and heat. These trichomes contain an oily resin which is packed with the beneficial phytocannabinoids that make the cannabis plant so valuable to our health and skincare. Depending on where and how the plant is grown, the phytochemical composition of these trichomes changes, and each strain will have a different therapeutic effect unique to it.

TERPENES

The same trichomes that release the potent phytochemicals from the cannabis plant also release another treasure, the highly aromatic compounds called terpenes. These little gems are also found in most essential oils (see page 32). They are responsible for the color and aroma of both essential oils and cannabis, and will substantially influence the medicinal and psychoactive effects of the plant and its extracted oil.

Aromatherapists and herbalists have applied the science of terpenes for years to affect the body and the mind. Highly varied, they have a multitude of applications, ranging from promoting relaxation and relieving stress to stimulating focus and clarity. They can add great power to a skincare formulation and offer additional support as they mediate the body's interaction with cannabinoids. Dr Ethan Russo has expanded on this theory, asserting that phytocannabinoids and terpene interactions enhance the therapeutic effects of cannabis. We are waiting for more research to come through to help us define how and to what extent, but early research supports the theory that terpenes enhance the effect of cannabinoids, which means they add great value to our beauty routine!

Below: CBD oil is extracted from the *Cannabis sativa* species of the plant, which is the name for both cannabis and hemp.

Plant materials

A cannabis material is a material—whether liquid, solid, or powder—which has been derived from the cannabis plant. Cannabis materials can be derived from the whole plant or an isolated part of the plant.

There are more and more cannabis and hemp materials available to us today, but this area can be a minefield. You will need to be able to decide which materials best suit your particular requirements, taking into account the legal environment where you live. Here we will look at how cannabinoids are extracted from the plant and what CBD materials are available to us.

Cannabis and hemp plants are most typically extracted by a process called CO_2 extraction or supercritical CO_2 extraction. This extraction method utilizes a chemical solvent such as ethanol, propane, butane, or hexane. The solvents are used to dissolve the plant into a solution or crude. The crude is then distilled in order to isolate certain components from the plant; phytocannabinoids, terpenes, flavonoids, chlorophyll, and even waxes can be extracted through chemical methods. This produces concentrates (also known as extracts). Once the extraction is complete, the solvents must be purged from the resinous oil. This can be achieved by employing evaporation, vacuuming, or hand-whipping. The aim is to remove the residual solvent from the extracted material to give us a clean, cannabinoid-rich material.

FULL SPECTRUM

Full spectrum material, also known as whole plant material, is a direct extract from the raw cannabis or hemp plants. It contains the full range of phytocannabinoids, terpenes, flavonoids, fatty acids, and other phytochemicals. While this range can vary according to the strain of the plant from which the material is extracted, full spectrum extract tends to have higher levels of phytocannabinoids and, in most cases, deliberately cultivated high levels of either THC or CBD—or even both. Full spectrum extract is most commonly used in medicinal and recreational products. All cannabinoids naturally found in the cannabis and hemp plant are present in this material. Full spectrum extract naturally contains terpenes, which amplify the effect of phytocannabinoids when interacting with the endocannabinoid system.

DISTILLATE

Distillate is also known as broad spectrum CBD, or CBD distillate. It is an extract which retains all the phytocannabinoids, terpenes, flavonoids, fatty acids, and phytochemicals of the original plant. This extract then goes through a special process after extraction to remove the THC cannabinoid. While this can vary according to the strain of the plant from which the material is extracted, distillate tends to have high levels of CBD, as well as a range of other naturally occurring cannabinoids. Distillate is considered ideal for people who want to experience the benefits of full spectrum without the THC content. It is used a lot in skincare and herbal products where brands or medicine makers don't want to take the risk of accidentally using illegal levels of THC. All cannabinoids naturally found in the cannabis and hemp plant are present in this material, with the exception of THC. Distillate naturally contains terpenes, which amplify the effect of phytocannabinoids when interacting with the endocannabinoid system.

ISOLATE

CBD isolate is CBD in its purest form. During the extraction process, all phytochemicals are removed or filtered out of the cannabis or hemp plant except for the CBD, so a CBD isolate contains no terpenes. Typically, a CBD isolate is between 90 and 97% CBD. The higher the percentage of CBD, the better.

Isolate is used in medicinal and recreational product preparations, but not as frequently as full spectrum and distillate. It is used a lot in beauty products, since there is no risk of violating the legal THC limit of 0.3%.

HEMP ESSENTIAL OIL

You may see a product called cannabis or hemp essential oil. It is important to understand that this admittedly lovely oil is not CBD oil. An "essential oil" is a distillation of plant material and this action captures the volatile components—for example, the aromatic terpenes. Cannabinoids are considered non-volatiles: they are fat loving, not water loving. This means that they cannot be distilled effectively and therefore cannot end up in the plant's essential oil.

A brief history of cannabis

Humankind discovered the medicinal properties of cannabis very early on! The earliest archaeological evidence for the use of cannabis is dated to 5,000–6,000 years ago in Central Asia and China. Prescriptions for cannabis in ancient Egypt include treatment for the eyes (glaucoma), inflammation, and cooling the uterus, as well as administering enemas. In ancient Greece, it was used in treating horses, especially for dressing sores and wounds, and treating inflammation, earache, and nosebleeds in humans. It took cannabis a very long time to reach western Europe—not until the 19th century.

CANNABIS IN NORTH AMERICA

The cannabis plant has had a long and controversial history in North America. Cannabis was likely brought there by the Spanish in the late 1500s. From its earliest cultivation by 17th-century settlers in the USA, the plant was used to create many products, including cloth, paper, sails, and rope. By the 1600s, farmers in colonies like Virginia, Massachusetts, and Connecticut were growing the plant. It was also used for its therapeutic properties—a use which continues with the groundbreaking medicines of today.

Until 1906, there was no such thing as an illegal drug across America. Cannabis, along with alcohol, morphine, opium, and other opiates, could be obtained freely at any pharmacy, without a prescription. From the early 20th century, a series of federal laws were introduced to regulate the use of cannabis and provide stricter law enforcement for its cultivation and possession.

In the 1970s, social attitudes began to change. There was an increase in popular support due to early research findings and anecdotal evidence of the medical benefits of cannabis. The first change began with decriminalization. In 1973 Oregon became the first state to decriminalize cannabis use, imposing a nominal fine ($100) for possession. More states followed, including California, Alaska, Mississippi, New York, North Carolina, and Colorado. In 1996, California became the first state to legalize cannabis for medical use.

Once states began to legalize medical cannabis, patients seeking treatment registered in the newly legalized states. 32,000 people registered as medical cannabis patients in Denver and at least 25,000 in Seattle when their states legalized cannabis.

A MARIJUANA MESSAGE

Generally, we don't use the term "marijuana" in the CBD beauty profession. After the Mexican Revolution in 1910, Mexican immigrants poured into US border states, bringing with them their favorite intoxicant, a plant they called *mariguana*. It subsequently became known as *marihuana* and *marijuana*. During the 19th century, the word "cannabis" was almost exclusively used to refer to the plant. However, when anti-Mexican sentiment in the United States began to rise in the early 20th century, the term was switched to "marijuana" to draw attention to the drug's use by Mexicans—and thereby attempt to convey a negative connotation. These terms were used in a derogatory way, and the association of the drug with Latin Americans had racial undertones as it grew in popularity in the 1920s among Jazz Age musicians and during the Prohibition Era.

The movement toward legislation is slowly but surely gaining momentum. Cannabis is now legal for all adults in 16 states of the USA, and for medical purposes in 35.

In Canada, cannabis was legalized in October 2018 with the Cannabis Act. The act permits possession, use, and cultivation in limited amounts for those over 18 years of age. Cannabis for medicinal use can only be purchased from federally licensed sellers.

Using CBD

CBD products come in many different formats. You can use CBD by applying an infused oil, cream, salve, or even transdermal patch. Transdermal delivery is a more specific term for a topical application that has been modified to increase absorption through the skin (in much the same way as a nicotine patch works).

Topical application

We know that topical CBD products interact with cannabinoid receptors in the skin and nervous system. Applying CBD oil to the skin is most efficacious when consistently applied directly to areas of chronic inflammation or pain, and if combined synergistically with other botanicals and essential oils to increase permeability. All the recipes in this book are for topical application.

Smoking or vaping

Traditionally, smoking was the most common way of consuming the plant. Today, cannabis flower or concentrate is often inhaled using a vape pen. Whether smoking or vaping, the cannabinoids are delivered directly to the blood and brain via the lungs. Terpenes are also volatilized and delivered directly to the brain through the same process. The effects of inhaled cannabinoids are experienced intensely and almost immediately.

Oral consumption

When cannabis is consumed orally, it has a very different effect compared to when it is smoked. It can take up to two hours to enter the bloodstream. However, once in the bloodstream, it tends to stay there for much longer than smoked cannabis—sometimes up to four to six hours, as opposed to only two hours when smoked. Orally consumed cannabis has a more potent effect and a longer duration, making it a great choice for night-time use, particularly in cases of chronic pain-related insomnia. Many oral CBD products are available as alcohol-based tinctures or edibles like gummies, chocolate, or candy. As with all orally consumed cannabis, there is a risk of taking too much and having an unpleasant experience. Since it can take up to two hours to feel the effects, you might be tempted to take an additional amount. If one's ideal dose is exceeded in this way, there can be an increase in side effects, including anxiety, paranoia, sleepiness, tachycardia (a heartbeat exceeding the normal rate), and even mild hallucinations.

Left: CBD oil is only used in small amounts in beauty recipes, but each drop is very powerful.

Your CBD Buying Guide

There are so many CBD products on the market today that you can get tricked into buying something that isn't the real deal. On pages 140–141 I have provided a list of tried and trusted suppliers. If you prefer to shop locally, follow these guidelines to help you choose.

LEGAL CONSIDERATIONS

Be sure you are familiar with the cannabis legislation where you live.

USE A REPUTABLE BRAND

When buying your CBD oil, it is important to choose reputable brands with a history of supplying plant-based concentrations.

CBD CONTENT

It may sound obvious, but you do need to check that the product you are purchasing actually contains CBD! A lot of stores are selling hempseed carrier oil as cannabis or CBD oil, which it is not (see page 16). Equally, beware of products labelled hemp essential oil—they do not contain CBD (see page 18).

LABEL INFORMATION

Look at the label! It should include all of the following information:

• Concentration of CBD in your product: While I have personally created products with up to 10,000 mg of CBD oil per 30 ml, I recommend a concentration of 300 mg of CBD oil per 30 ml for the recipes in this book.

• Batch number: You should be able to see a batch number on the label to ensure the product is traceable to the manufacturer in case of any problems with it.

• Best-by date: This is vital so that you can gauge the freshness of the product and ensure you use it before it expires.

Above: **Choose the best-quality CBD oil for your recipes, such as the Red Raspberry Seed Facial Oil (see page 82).**

LAB REPORT/CERTIFICATE OF ANALYSIS

It is very important to look for a Certificate of Analysis (COA) for the product you are interested in. This should be readily available on the brand's website. It is important to know if the analysis was performed by an accredited laboratory. A good indication will be if they are accredited in accordance with the International Organization for Standardization (ISO). This report will show you the CBD concentration. It is a nice comfort check to ensure your product truly has the CBD concentration advertised. Note that the lab report should be reasonably recent, preferably within the last 12 months.

CHAPTER 2

How the Skin Works

When Mother Nature created skin, she provided us with the most miraculous packaging! Skin protects our internal organs; it is our first line of defense against viruses, germs, pollutants, pathogens, harmful chemicals, and environmental stresses. It insulates our bodies, regulates temperature, prevents dehydration, and synthesizes Vitamin D from sunlight. It is elastic, flexible, and durable. A myriad of sensory fibers and nerve endings present in skin create an important early warning system when danger threatens, prompting the skin to mobilize the immune system to defend the body. All the sensations we experience—whether pleasurable or painful—are provided by the skin.

Since everything you will create using the recipes in this book will be applied to and absorbed by your skin, let's take a good look at this marvelous organ, which happens to be the largest organ in our bodies.

The Skin's Anatomy

What happens when CBD and skincare botanicals are applied to our skin? To answer this question, you need first to know a little about the skin's structure.

The skin consists of three main layers. The surface layer is called the epidermis (this word comes from the ancient Greek word *epi*, meaning "on top of," and *derma*, meaning "skin"). The second main layer is the dermis. The bottom layer is called the hypodermis (from the Greek word *hypo*, meaning "beneath").

The epidermis

The epidermis is the surface of our skin. It varies in thickness from 0.05 mm on our eyelids to up to 1.5 mm on the soles of our feet. The primary function of the epidermis is to act as a barrier, preventing water loss and, importantly, blocking the entry of pathogens. It is equipped with a sophisticated defense system which triggers our immune response when it encounters microbial and bacterial threats. It also protects our skin against environmental stresses, harmful chemicals, and ultraviolet radiation (by producing melanin, which shields the body against sun damage and gives our skin its color).

In addition, the epidermis is a cell production factory, constantly generating new cells and sending them up to the surface to replace dead cells. These shed so rapidly that the skin will completely replenish itself every 28 days or so. It is worth noting that the natural shedding process can be slower in aging skin, hence the value of regular exfoliation. Exfoliation is a powerful practice for removing dead cells more quickly than they would be shed naturally. This action allows new cells to rise to the surface of the epidermis to offer a brighter, more youthful complexion. Aside from the skin benefits of exfoliation, it will also help increase circulation and improve lymphatic drainage, giving our skin a more radiant glow. We provide some wonderful exfoliating recipes in this book!

The dermis

The dermis lies immediately underneath the epidermis. Its main role is to support the epidermis, providing nourishment to the surface of the skin. This layer of the skin is a real workhorse! It produces sensation, sweat, hair, and oil (to keep your skin soft); it also carries out the vital task of bringing blood to your skin.

The dermis is rich in collagen and elastin—collagen gives skin its toughness and strength, and elastin, as the name suggests, provides elasticity and resilience. Together, they provide structure to the skin and are very important for healing wounds. However, both are affected by increasing age and exposure to sunlight, which result in sagging and stretching of the skin.

The hypodermis

The hypodermis contains most of the fat that we worry about when we want to lose weight. However, this layer of fat has many useful functions! It acts as a shock absorber, helping protect our body from injury. It is also a natural insulator, protecting our body from heat loss. Finally, it provides a valuable energy reserve for the body to call on when needed.

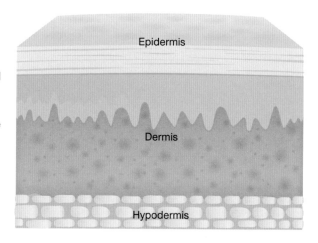

Right: **The three main layers of the skin.**

The natural moisturizing factor

Did you know that your skin has its own amazing inbuilt moisturizer? It is called the natural moisturizing factor, usually abbreviated as NMF. The NMF is not a single magic bullet, but a mix of powerful water-loving substances in the skin that keep it protected and hydrated. Together, they draw moisture like a magnet into the epidermis, the surface of the skin.

However, many factors can impede the efficiency of this extraordinary system. For example, in an atmosphere where there is no humidity, our skin will easily become dehydrated. Sun exposure, cold and windy weather, poor diet, and air conditioning can all have an adverse impact on our skin.

Unfortunately, some common cosmetic ingredients can have the same effect. One of the main offenders is an ingredient called a surfactant. This word is shorthand for "surface acting agent." It is a very important ingredient in a host of cleaning, foaming, and emulsifying products. What makes surfactants useful is their ability to combine otherwise incompatible materials—like water, oils, fats, and solvents. They are a key ingredient in some cosmetics, especially facial cleansers, body washes, and shampoos.

Above left: **Shea butter and cocoa butter are excellent moisturizers and are used in the Vanilla-enriched Body Butter (see page 100).**

Above: **Vegetable oils (see page 40) provide another way of getting moisture into the skin, as used in the Frankincense Nourishing Body Oil (see page 102).**

However, many surfactants strip the terrific natural barrier that is the NMF, resulting in dry, irritated, flaky, and distressed skin. Watch out especially for products containing sodium lauryl sulfate. In this book you will find many beautiful, surfactant-free alternatives for cleaning, toning, nourishing, and moisturizing your skin—all of which you can easily make in your own kitchen!

Treating Skin Types

We all have different skin types. Our skin produces sebum, an oily, waxy substance, to protect it. The amount of sebum produced in skin helps to regulate the effectiveness of the skin's barrier function and therefore the condition of your skin. Too much sebum can lead to oily skin, whereas low sebum production can cause dry skin.

We can even experience various skin types and conditions throughout our lives, depending on our lifestyle choices and what is happening within our bodies. Hormones will dramatically affect our skin—for example, during pregnancy, your skin may carry the "pregnancy glow" or you may experience hormonal acne. Menopause is another example—as you transition through the menopause, your skin can become very dry or even dehydrated as your estrogen levels deplete. Stress and environmental factors will affect your skin type—for example, living in a city will subject your skin to more harsh pollutants, and living in a sunny country will increase your risk of experiencing a reduction in elasticity and possible premature aging.

Remember, regardless of your skin type, your largest organ deserves tender, loving care. The delectable and replenishing beauty recipes in this book should be supported by a diet rich in skin-nourishing nutrients, along with regular exercise and restful sleep!

Dry skin

The key to treating dry skin is to use ingredients which feed the skin's outermost layer, its lipid barrier, keeping it hydrated and infused with essential fatty acids (see page 29). Certain vegetable oils, such as hempseed and peach kernel oils (see pages 43 and 44), are an ideal skincare ingredient which will deliver a powerful fatty-acid boost to dry or aging skin, slowing the onset of inflammation and the formation of fine lines and wrinkles.

Oily skin

One of the age-old myths of skincare is that oily skin needs to be aggressively cleansed. The irony is that "squeaky clean" skin is skin that has been stripped of its natural protection and is thus more susceptible to bacterial overgrowth. One of the first lessons in chemistry classes is that "oil dissolves oil," and it is particularly relevant to skincare. The kindest and most effective way to clean oily skin and balance an overproduction of sebum is to use natural oils. A light oil such as argan (see page 42) will cleanse the skin without disturbing its pH balance. It will also neutralize the bacteria that lead to unwelcome breakouts.

Sensitive skin

The focus when treating sensitive and easily irritated skin is to strengthen the skin's barrier function. The recipes in this book harness the robust anti-inflammatory and antimicrobial ingredients of CBD, supported by other ingredients, to calm, soothe, support, and protect the skin.

Right: **By identifying your skin type, you can choose the most appropriate products for your skin.**

Vitamins, Minerals, and Other Nutrients

Skin is waterproof but, contradictory though it sounds, it is also selectively permeable, allowing some fat-soluble substances, including vitamins and minerals, to pass through the epidermis. We have all heard of vitamins and minerals and know that they help our body—but what are they, and how can they help our skin?

Vitamins

These are compounds we need for our bodies to work properly. There are a range of vitamins, each with its own benefit to the skin.

- Vitamin A is supportive to the skin and has the ability to produce new skin cells.

- Vitamin B, including B_1, B_2, and B_6, is soothing to the skin and promotes skin cell turnover, which helps to replenish the skin's complexion.

- Vitamin C supports the skin's immune function, is necessary for collagen production, and protects against signs of aging.

- Vitamin D is known to repair the skin while also stimulating skin cell growth, which results in a rejuvenated complexion.

- Vitamin E is an antioxidant which protects the skin against the effects of free-radical damage (see page 41). It can be bought in liquid form to add to your products (see page 60).

- Vitamin K is anti-inflammatory to the skin and supports the production of collagen.

Minerals

Essential minerals have a role in maintaining skin development and function while working to optimize the overall health of the skin cells.

- Magnesium is a mineral which supports our skin, hair, and muscles. For the skin, magnesium supports the

Above: **Raspberries and blueberries are rich in vitamins and can be used in your homemade CBD beauty recipes.**

synthesis of collagen while also working to promote healthy hair follicle growth. For the muscles, it has the ability to ease muscular pain and tension.

- Zinc is soothing and anti-inflammatory for our skin. It can also protect our skin against damage by free radicals.

- Calcium is antioxidant to the skin and a natural skin moisturizer, owing to its ability to regulate our skin's production of sebum.

- Sulfur is antimicrobial to the skin, which results in cleaner, fresher skin. It also acts like an exfoliator, helping to shed dead skin cells while encouraging new skin to come through, so offers a fresh complexion.

- Sodium is tonifying and cleansing for the skin and helps to restore your skin's natural protective barrier.

- Potassium works to regulate the moisture levels in your skin cells and can act as a hydrating agent for your skin.

- Chromium is known to cleanse acne-prone skin.

- Iron is known to promote a healthy glow in the skin.

Essential fatty acids

Now, "fatty acids" may sound off-putting. Many people mistakenly think that to maintain a healthy lifestyle they must avoid eating fats. The fact is that your body actually needs certain healthy fats to function properly. "Good" fats build cell membranes, insulate nerves, and ensure that many vitamins, including A, D, E, and K, work the way they should. These are called essential fatty acids (EFAs) because your body cannot produce them on its own. They must be acquired in your diet, though it is now increasingly recognized that they can also be acquired through the skin. And this is what interests us in this book! EFAs abound in good things and are present in every vegetable oil mentioned in this book.

There are three main categories of fatty acids: saturated, monounsaturated, and polyunsaturated.

SATURATED FATTY ACIDS

Saturated fatty acids are antimicrobial, moisturizing, rejuvenating, and hydrating, especially to hair. Many saturated fatty acids are anti-inflammatory and are ideal when treating mature skin as they hydrate and plump out the skin. Another important benefit is that they form healthy cell membranes, giving us a fresher complexion. Coconut oil is an example of an oil that is rich in saturated fatty acids. Saturated fatty acids include caproic acid, caprylic acid, capric acid, lauric acid, myristic acid, palmitic acid, and stearic acid.

MONOUNSATURATED FATTY ACIDS

Monounsaturated fatty acids are primarily protective to the skin as they reduce oxidative stress, protect against damage from the sun, maintain moisture levels in the epidermis (see page 24), and are powerfully anti-inflammatory. You will find them in oils such as almond, peach, olive, and avocado. Monounsaturated fatty acids include oleic acid, erucic acid, and eicosenoic acid.

Left: **Shea butter is rich in fatty acids and vitamins, so the Shea and Apricot Facial Cream (see page 84) is an excellent moisturizer.**

POLYUNSATURATED FATTY ACIDS

Polyunsaturated fatty acids hydrate and moisturize the skin and protect the skin's natural barrier. They also help fight free radicals (see page 41). You will find them in oils such as grapefruit, hempseed, and rosehip. Polyunsaturated fatty acids include linoleic acid, alpha-linolenic acid, gamma-linolenic acid, and eicosadienoic acid.

OMEGAS

These are classified as polyunsaturated fatty acids and all carry unique therapeutic properties.

- Omega-3 is soothing to dry and irritated skin. It will also regulate the skin's own oil production, helping to improve hydration in the skin and minimize signs of aging.

- Omega-6 is soothing to the skin and works to protects the skin's natural barrier.

- Omega-9 is hydrating and soothing to the skin.

Other nutrients

There are other beneficial nutrients that can be found in natural ingredients.

- Squalene is an antioxidant and has the ability to boost collagen production within the skin. It is also hydrating and works to fight free-radical damage.

- Beta-carotene is a precursor of vitamin A and is packed with antioxidant power, which will protect our skin against free-radical damage and prevent premature aging.

- Tocopherols are antioxidants which are found in many foods, including tomatoes and red raspberries.

- Amino acids hydrate, moisturize, and soothe the skin. The antioxidants provided by amino acids will help to protect the skin from free-radical damage.

CHAPTER 3

Skin-Loving Botanicals

In the following pages, I will walk you through the cornucopia of natural ingredients in the recipes in Chapters 5–8 so that you can make informed decisions about the best ingredients for your skin. The 35 recipes are all nut-free and vegan, and I recommend that you choose natural, organic ingredients where possible. You are in for a treat!

Essential Oils

Essential oils have been my life's work. I have spent the last 20 years learning about, loving, and sharing the chemistry and clinical use of these formidable plant essences, and what a gift it has been! I never cease to marvel at these astonishing, complex, and beneficial botanicals which have been used since ancient times for their aromatic and therapeutic properties. Now, with the prevailing enthusiasm for green and vegan beauty and wellbeing products, there is a new focus on these 100% natural skincare ingredients.

What are essential oils?

Essential oils are the distilled aromatic essence of a plant. They are made by steaming or cold pressing various parts of a plant (flowers, bark, leaves, or fruit) to capture the compounds that produce fragrance. Plants are packed with phytochemicals. These are a powerful group of nutritive and antioxidant compounds that also provide plants with their color, flavor, and aroma. Each essential oil consists of hundreds of these phytochemicals, all complementing each other. A single oil is frequently a multitasker. Lavender essential oil can relieve skin irritation, soothe muscular discomfort, and calm nervous tension in one application. Rosemary will act simultaneously to clear blocked sinuses, stimulate hair growth, gently settle nervous anxiety, and promote mental clarity and focus.

Essential oils are individually unique in so many ways, even down to their color, their viscosity, and the amount of plant material required to make the oils. You would need 60 roses to acquire one drop of rose essential oil, whereas the rind of five lemons will yield 20 drops of lemon essential oil.

Essential oils will strengthen and calm your mind, nourish your soul, stimulate sensuality, support your immunity, love your skin, heal your body, and balance your mood. Their range and versatility are breathtaking.

Botanical scent

The sense of smell is the most primal of our senses. When essential oils are inhaled, they go directly to the brain and the impact of their therapeutic properties is almost immediate. For example, if we inhale an essential oil with sedating properties, such as ylang ylang, it will produce a relaxing response very rapidly.

Essential oils broadly fall into one of the following types of aroma—herbal, citrus, spicy, woody, resinous, fruity, earthy, or floral. However, there are layers of subtlety within each of these categories, leading to depths of

Right: Lavender essential oil has a whole host of benefits, including relieving muscular pain and tension headaches.

aromatic complexity unique to each individual essential oil. The aroma of an essential oil tends to be determined by the part of the plant from which it is extracted. Whether the scent is of grapefruit peel, lavender flowers, cinnamon bark, bay laurel leaves, or rose petals, they will have one thing in common—they will all smell heavenly!

There is a very convenient scent classification method to help you identify the aroma of your essential oil. An essential oil can be a top, middle, or bottom note. Essential oils which are top notes will be energizing and uplifting—for example, citrus oils, like orange and grapefruit. Floral essential oils, like geranium and lavender, tend to be middle notes, offering a balancing aroma. The more resinous oils, like frankincense and Palo Santo, are base notes, having the ability to calm us and support us in meditation.

Extraction Methods

Steam distillation is the most common method of extraction for essential oils. The plant material is placed in a still, and steam is passed through it, releasing the plant's aromatic oil. The oil-infused steam goes to the top of the still, where it is cooled in a condensing chamber and turns back into water. As oil is lighter than water, it floats on top and is easily removed.

Cold pressing, also known as **expression**, is the most common and effective way of extracting essential oil from the rind of various citrus fruits. The skin is punctured using a spiked press, releasing the aromatic oils. The fruit is then sprayed with water and the resulting mixture filtered and centrifuged. Since no heat is involved, the result is a particularly potent and fragrant aroma.

Solvent extraction is used on the most precious flowers which are too fragile to undergo distillation. This method uses food-grade solvents, such as hexane and ethanol, to isolate the phytochemicals from the plant material. Solvent extraction often yields what we call absolutes or concentrates, which are commonly used for perfume making. Plant oils, such as rose and jasmine, are typically obtained via solvent extraction.

CO_2 extraction has become a popular method of extraction in the last few years. It uses carbon dioxide to extract the phytochemicals from the plant material. While this method of extraction is common in the cannabis world, it is less frequently used in aromatherapy.

Above: **There are many stores selling essential oils—always buy from a known and trusted brand.**

Essential oils and CBD

Some of the most common compounds in essential oils and in CBD oil are terpenes (see page 17). Each terpene is associated with unique effects, typically supported with scientific research or/and anecdotal evidence. Aromatherapists and herbalists have used the science of terpenes for years when formulating plant-based blends to create certain desired effects on the body and mind, whether to promote relaxation or to stimulate focus and clarity, plus many more effects.

The chemistry of CBD oil and essential oils complement each other, creating a synergy in which the whole is greater than the sum of its parts. The differences between the oils can be subtle, but essential oil and CBD terpenes add great power when combined. Skin permeability, the properties of the oils, and their therapeutic value are all enhanced by this dynamic pairing.

Essential oil safety

The hundreds of powerful phytochemicals in essential oils are easily absorbed by the skin. For this reason, these volatile and highly concentrated plant essences need to be used wisely and with care. They may be our powerful allies when treating and supporting our body and mind, but, owing to their concentration, they can have unwanted effects, such as skin irritation, sensitization, and phototoxicity (when the skin is ultra-sensitive to sunlight and UV rays). Therefore, they should always be diluted in a vegetable carrier oil (see page 40). Diluting also means that you will use less of your precious essential oil without any loss of efficiency.

You should never attempt to use essential oils undiluted or internally. This is only to be done by clinical aromatherapists, who have the knowledge, training, and experience to do so safely.

Buying and storing essential oils

When buying essential oils, there are a few important things to look out for:

• The oil should come in a dark glass bottle, which prevents exposure of the oil to sunlight (see below).

• The label should include the Latin name of the plant (e.g. Lavender, *Lavandula angustifolia*), the country of origin, and a batch number.

• You should buy from a reputable company that has been around for some years. You will find a list of tried and trusted suppliers on page 140.

Oils vary greatly in their shelf life. Citrus oils will evaporate rapidly if exposed to air; even when looked after meticulously, they will need to be used up within one year. Resinous oils such as myrrh get richer with age, just like a fine red wine.

They may be called "essential oils," but these botanical extracts are not oily as such, they are volatile compounds which will evaporate easily if exposed to direct sunlight or heat.

Tread lightly

With demand for essential oils on the increase, we need to be aware of endangered plant species. Certain plants need safeguarding, and some are undergoing extensive replanting programs to preserve them. The best way to ensure that you do not inadvertently buy oils derived from vulnerable plants is to go for products that come from reputable producers.

When using essential oils, the "less is more" theory is not only relevant, but it should also be engaged continually! These potent botanicals are highly effective in small doses, so let's not be wasteful of the gifts Mother Nature has bestowed on us.

Finally, our planet is precious and deserves more respect than she has received from us mere mortals who walk through her. We are only here for a short time while she works to care for future generations. Let us be conscious of that, and strive to use her gifts with kindness, wonder, and appreciation.

Right: **Essential oils are usually stored in dark, glass bottles (see page 60).**

Your Guide to Classic Essential Oils

Some of the botanicals we cover in this section will be familiar to you and some will be thrillingly new. This detailed ingredient chart describes the scent, aroma, viscosity, and cosmetic and therapeutic properties of some of my favorite essential oils, along with relevant safety precautions. You will also learn the common and botanical names of each oil, along with its extraction method. Armed with this information, you will be able to don your apron with confidence as you step into your kitchen and start creating your very own CBD beauty skincare products!

BOTANICAL INFORMATION	METHOD OF EXTRACTION	THERAPEUTIC PROPERTIES		
		FOR THE SKIN	FOR THE BODY	FOR THE EMOTIONS
Black pepper *Piper nigrum* Scent note: middle Aroma: spicy, warm, and radiant Safety precautions: nontoxic, nonirritant when diluted	Steam-distilled from the dried fruit	Regenerative, anti-inflammatory, antibacterial, and analgesic.	Anti-inflammatory to muscular and joint pain and stiffness, digestive aid, and supports the circulatory system.	Stimulating, energizing, and encouraging.
Carrot seed *Daucus carota* Scent note: middle Aroma: earthy, herbaceous, and warm Safety precautions: nontoxic, nonirritant when diluted.	Steam-distilled from the dry seeds	Rejuvenating to mature skin, antioxidant, anti-aging, tonifying, moisturizing, and wound healing. Helps when treating eczema, psoriasis, skin ulcers, and skin infections.	Digestive aid, eases menstrual cramps, and cleansing to the liver.	Energizing, calming, and relaxing.
Cedarwood *Cedrus atlantica* Scent note: base Aroma: woody and earthy Safety precautions: nontoxic, nonirritant when diluted. Avoid topical use during pregnancy or while breastfeeding.	Steam-distilled from the wood of the tree	Anti-inflammatory, soothes irritation, skin toning, anti-acne, and antifungal.	Eases arthritic pain, respiratory decongestant, and eases nervous tension.	Grounding, calming, and uplifting.

BOTANICAL INFORMATION	METHOD OF EXTRACTION	THERAPEUTIC PROPERTIES		
		FOR THE SKIN	FOR THE BODY	FOR THE EMOTIONS
Frankincense *Boswellia carterii* Scent note: base Aroma: resinous, warm, and spicy Safety precautions: nontoxic, nonirritant when diluted	Steam-distilled from the resin of the tree	Wound healing, moisturizing, regenerating, rejuvenating to mature skin, helps to revive scar tissue, and anti-inflammatory.	Pain relieving (especially arthritic pain), eases muscle stiffness and ache, relieves nervous exhaustion, supports sleep, and supports immunity.	Calming, stress relieving, and eases anxiety.
Geranium *Pelargonium graveolens* Scent note: middle Aroma: herbal, sweet, rosy, and lemony Safety precautions: nontoxic, nonirritant when diluted.	Steam-distilled from the leaves of the plant	Regenerating, moisturizing, balances sebum production, hydrates dry skin, and promotes healthy hair.	Eases the symptoms of premenstrual tension and eases menopausal-induced mood swings and hot flushes.	Calming, balancing, nourishing, and stabilizing when feeling anxious.
Ginger *Zingiber officinale* Scent note: middle Aroma: spicy, sweet, and radiant Safety precautions: nontoxic, nonirritant when diluted.	Steam-distilled from the rhizomes of the flowering plant	Promotes healing of bruises, and revitalizing.	Soothes digestive upset, antispasmodic, relieves joint and muscle pain and stiffness, and improves circulation.	Warming, energizing, restores motivation, and balancing to emotions.
Grapefruit *Citrus paradisi* Scent note: top Aroma: refreshing and citrus Safety precautions: nontoxic, nonirritant when diluted. Avoid exposure to the sun for 12 hours after use.	Cold-pressed from the rind	Detoxifying, anti-acne, antimicrobial, and cleans congested skin.	Eases migraines, stimulates lymphatic circulation, tonifying, diuretic, and eases muscular ache.	Uplifting, calming, comforting and balancing, and promotes positive energy.
Lavender *Lavandula angustifolia* Scent note: middle Aroma: floral, fresh, and relaxing Safety precautions: nontoxic, nonirritant when diluted.	Steam-distilled from the flowers	Anti-inflammatory and useful for treating burns, bug bites, wounds, bee stings, rashes, acne, and skin irritation.	Supports the immune system, relieves muscular and joint aches and pains, soothing to the nervous system, relieves tension headaches, supports sleep, and alleviates menstrual cramps.	Eases emotional tension and invites happiness and positive energy.
Lemon *Citrus limon* Scent note: top Aroma: citrus, fresh, and lemony Safety precautions: nontoxic, nonirritant when diluted. Avoid exposure to the sun for 12 hours after use.	Cold-pressed from the rind	Cleansing, detoxifying, and antiviral.	Decongestant and supports the immune and circulatory systems.	Energizing and aids concentration.

BOTANICAL INFORMATION	METHOD OF EXTRACTION	THERAPEUTIC PROPERTIES		
		FOR THE SKIN	FOR THE BODY	FOR THE EMOTIONS
Marjoram *Origanum majorana* Scent note: middle Aroma: herbaceous, warm, and sweet Safety precautions: nontoxic, nonirritant when diluted. Not to be used during pregnancy or while breastfeeding, or on children under 10 years.	Steam-distilled from the leaves and flowers	Promotes healing of bruises, cleansing, antibacterial, and antifungal.	Provides a range of effective digestive supports and supports the respiratory system, improves blood circulation, eases arthritic pain (especially rheumatism), relieves nervous tension, and eases menstrual cramps.	Fortifying, stress-reducing, emotionally supportive, and relieves insomnia.
Orange *Citrus sinensis* Scent note: top Aroma: citrus, fruity, fresh, and sweet Safety precautions: nontoxic, nonirritant when diluted.	Cold-pressed from the rind	Tonifying, revitalizing to lackluster skin, anti-inflammatory, and promotes wound healing.	Restorative, antispasmodic, and detoxifying.	Combats fatigue, inspires positive energy, and invites joy and happiness.
Palo Santo *Bursera graveolens* Scent note: base Aroma: resinous, warm, and radiant Safety precautions: nontoxic, nonirritant when diluted.	Steam-distilled from the wood of the tree	Nourishing, moisturizing, anti-acne, anti-inflammatory, and antioxidant.	Decongestant, eases muscular pain and restricted joint movement, and supportive to the respiratory system.	Calming and relaxing, so it assists with meditation and encourages concentration, and was used by the Incas to cleanse the spirit of negative energy.
Peppermint *Mentha x piperita* Scent note: top Aroma: minty, camphor-like, and fresh. Safety precautions: nontoxic, nonirritant when diluted. Not to be used during pregnancy or while breastfeeding. Do not use near the nose and face of children under 10 years.	Steam-distilled from the leaves	Cooling, analgesic, anti-inflammatory, antioxidant, detoxifying, and a skin decongestant.	Respiratory decongestant, anti-nausea, and eases headaches.	Mentally stimulating, and awakens and refreshes the senses.
Petitgrain *Citrus aurantium* var. *amara* Scent note: top Aroma: citrus, floral, and woody Safety precautions: nontoxic, nonirritant when diluted.	Steam-distilled from the leaves and twigs of the bitter orange tree	Regulates sebum production, regenerative, anti-acne, anti-inflammatory, soothing, and wound healing.	Supports sleep, stimulates the immune system, reduces muscle spasms, and eases nervous indigestion.	Calming and reassuring in times of anxiety, creates a balancing and uplifting effect, and supports emotional healing.

BOTANICAL INFORMATION	METHOD OF EXTRACTION	THERAPEUTIC PROPERTIES		
		FOR THE SKIN	FOR THE BODY	FOR THE EMOTIONS
Roman chamomile *Anthemis nobilis* (also known as *Chamaemelum nobile*) Scent note: middle Aroma: fruity, sweet, and warm Safety precautions: nontoxic, nonirritant when diluted. Safe to use with babies, children, and the elderly.	Steam-distilled from the growing plant	Hydrates dry and sensitive skin, promotes wound and scar healing, analgesic, anti-allergen, and anti-inflammatory.	Supports sleep.	Grounding, calming, and conducive to sleep.
Rosemary *Rosmarinus officinalis* Scent note: top Aroma: herbaceous, camphor-like, and fresh Safety precautions: nontoxic, nonirritant when diluted. Not to be used during pregnancy, while breastfeeding, or when suffering from epilepsy or high blood pressure. Do not use near the nose and face of children under 10 years.	Steam-distilled from the herb	Cleansing and antimicrobial.	Detoxifying, decongesting, anti-inflammatory, relieves headaches, and reduces muscular aches and pain (especially relating to arthritis).	Regenerative, energizing, and encourages mental clarity.
Tea tree *Melaleuca alternifolia* Scent note: middle Aroma: medicinal and camphor-like Safety precautions: nontoxic, nonirritant when diluted.	Steam-distilled from the leaves, twigs, and bark	Promotes healing of wounds, bug bites, shingles, boils, and blisters, cleansing and purifying to skin and nails, anti-acne, anti-inflammatory, antibacterial, and antifungal.	Decongestant and supportive to the respiratory system, boosts the immune system, and an ideal hand sanitizer.	Uplifting and antidepressant.
Ylang ylang *Cananga odorata* Scent note: base Aroma: floral, sweet, exotic, and sensual Safety precautions: nontoxic, nonirritant when diluted. Do not use on children under 10 years.	Steam-distilled from the flower	Cleansing, tonifying, anti-inflammatory, and soothes irritated skin.	Supports sleep, eases nervous tension, and supports reproductive health.	Relaxing, calming, encourages self-confidence, invites creativity, sensual, aphrodisiac, and eases tension and anxiety.

Vegetable Oils

Vegetable oils are oils with little or no scent which are used to dilute highly concentrated essential oils and "carry" them into the skin—for this reason, they are also known as carrier oils. You might think that diluting essential oils would be undesirable, but that is not the case. Essential oils are volatile, meaning that they evaporate quickly. Vegetable oils, on the other hand, are very stable in comparison. When delivered in a carrier oil, the essential oil is better absorbed into the skin.

Another consideration when applying essential oils topically is that some can be irritating to the skin, creating burning or itching sensations. This does not happen when the highly potent essential oil is dissolved in a neutral carrier oil—opening up the huge and diverse range of essential oils to all, regardless of skin sensitivity.

There is much, much more to vegetable oils than simply serving as a vehicle for their more glamorous passengers. They are magnificently therapeutic in their own right, complementing, enhancing, and extending the benefits of the aromatic oils that they carry.

Oils such as rosehip, red raspberry, and borage have been used for years for their regenerative, anti-inflammatory, and moisturizing qualities. Other oils, such as apricot and sesame oil, are used in wound care and treatment of sun damage. Still others, such as baobab and pomegranate, are rich in antioxidants and deeply hydrating and boost collagen production. All are rich in essential fatty acids (see page 29).

Vegetable oils and free radicals

One of the main contributors to aging and stressed skin is the presence of free radicals, which are produced by our bodies as a reaction to environmental stresses such as air pollution and ultraviolet (UV) radiation. A free radical is an abnormal skin molecule whose naturally paired electrons have been split. The incomplete molecule then seeks to stabilize itself by stealing an electron from a healthy cell. The healthy cell, now damaged, searches for another electron—thus damaging another cell, and a destructive chain reaction starts. An imbalance between free radicals and the body's own naturally produced antioxidants is called oxidative stress.

The good news is that cellular damage can be fought using natural vegetable oils. Vegetable oils contain antioxidants, which work to increase the skin's defenses against UV radiation and help maintain skin health and a clear complexion (the Probiotic Berry Mask recipe on page 74 is particularly effective in fighting free radicals).

The fatty acids comedogenic rating

An excellent way to understand how vegetable oils interact with your skin and especially your pores is what is called the comedogenic rating. Comedo is the medical word for a blackhead. The comedogenic rating measures the likelihood of a particular oil, butter, or emollient clogging your pores. The scale runs from 0–5:

0: Will not clog pores (often described as non-comedogenic)
1: Low probability
2: Moderately low
3: Moderate
4: Fairly high
5: High probability of clogging pores

Left: **There are a whole range of oils that can be used in your beauty recipes.**

Above: **Not only a delicious cooking oil, olive oil can repair sun-damaged skin cells and help prevent premature aging.**

Buying and storing vegetable oils

When purchasing, look for pure, cold-pressed, unrefined oils with no additives, and preferably organic. Most oils on supermarket shelves are extracted using chemical solvents, usually a petroleum distillate, to release the fat in the seed. The solvent is then cleaned off by heating the oil to 212°F (100°C). Then, to remove any lingering odor of solvent, the oil is bleached. It is a fast, efficient, and cheap process. Cold-pressed oil is produced by passing the seed or fruit through a hydraulic press to release the oil. That's it. No heat is applied. What you have is golden magic—the unrefined vegetable oil.

Store in a dark, cool place or, for oils with a shelf life of less than a year, refrigerate to extend the lifespan.

Your Guide to Vegetable Oils

The following chart covers all the vegetable oils that appear in the recipes
in this book, with information about their extraction method, nutrient content,
therapeutic properties, shelf life, and comedogenic rating, making it easy
to choose the best oil for your needs.

BOTANICAL INFORMATION	METHOD OF EXTRACTION	NUTRIENT CONTENT	THERAPEUTIC PROPERTIES
Apricot oil *Prunus armeniaca* Comedogenic rating: 2 Viscosity: light to medium Shelf life: 1 year Safety precautions: none	Cold-pressed from the kernel	Rich in vitamins E and K.	Nourishing, revitalizing, emollient, wound healing, anti-inflammatory, antioxidant, and hydrating. Soothes irritation, provides a protective and balancing barrier against environmental damage, and has good skin penetration properties.
Argan oil *Argania spinosa* Comedogenic rating: 0 Viscosity: light to medium Shelf life: 1–2 years Safety precautions: none	Cold-pressed from the kernel	Rich in vitamin E and squalene.	Protects against free-radical damage, anti-aging, anti-inflammatory, moisturizing, nourishing, wound healing, and antioxidant. Provides a protective and balancing barrier and has good skin penetration properties.
Arnica oil *Arnica officinalis* Comedogenic rating: depends on the infused base Viscosity: light to medium Shelf life: 6–9 months Safety precautions: nontoxic, nonirritant when diluted.	Cold-pressed from the flowers	Vitamin content depends on the infused base.	Antimicrobial, anti-inflammatory, analgesic, toning, eases acute inflamed conditions of muscular and skeletal tissue, and healing to scar tissue and bruised skin.
Baobab oil *Adansonia digitata* Comedogenic rating: 2 Viscosity: light to medium Shelf life: 2–4 years Safety precautions: none	Cold-pressed from the seeds	Rich in vitamins A, D, and E.	Regenerative, moisturizing, toning, skin softening, and luxurious. Relieves painful skin conditions, promotes wound healing, supports mature skin by improving elasticity and supporting collagen health. Effective when treating dehydrated, chapped, or flaky skin, and improves the appearance of stretch marks.
Black cumin oil *Nigella sativa* Comedogenic rating: 2 Viscosity: light to medium Shelf life: 3–4 years Safety precautions: none	Cold-pressed from the seeds	Rich in vitamin C.	Antibacterial, antiviral, anti-inflammatory, regenerative, and moisturizing. Improves elasticity and is effective when treating dry skin, wounds, acne, eczema, and psoriasis. Good skin penetration properties and nourishing to hair.

BOTANICAL INFORMATION	METHOD OF EXTRACTION	NUTRIENT CONTENT	THERAPEUTIC PROPERTIES
Borage oil *Borago officinalis* Comedogenic rating: 2 Viscosity: light to medium Shelf life: 6–12 months Safety precautions: none	Cold-pressed from the seeds	Rich in vitamin E.	Anti-inflammatory, antioxidant, improves elasticity, regenerative, and relieving to psoriasis, eczema, acne, and minor wounds. Reduces the appearance of fine lines, improves skin elasticity, and helps protect the skin from photodamage.
Castor oil *Ricinus communis* Comedogenic rating: 1 Viscosity: thick Shelf life: 1–2 years Safety precautions: none	Cold-pressed from the seeds	Rich in vitamin E.	Anti-inflammatory, hydrating, and soothing. Provides a protective barrier to the skin, helps minimize trauma to the skin, and has good skin penetration properties.
Coconut oil *Cocos nucifera* Comedogenic rating: 4 Viscosity: thick Shelf life: 2–3 years Safety precautions: none	Cold-pressed and centrifuged	Rich in vitamins A, D, and K.	Anti-inflammatory, emollient, anti-aging, antimicrobial, antibacterial, and antifungal. Fights damage by free radicals, supports skin's immunity, and has an affinity for hair protein and the capacity to penetrate the hair shaft, moisturizing and softening it.
Comfrey oil *Symphytum officinale* Comedogenic rating: depends on the infused base Viscosity: medium Shelf life: 2 years Safety precautions: nontoxic, nonirritant when diluted.	Cold-pressed from the leaves	Rich in vitamins A, C, and E.	Anti-inflammatory, astringent, analgesic, and antifungal. Eases swelling and muscular skeletal pain, and is healing to minor burns and rashes.
Hempseed oil *Cannabis sativa* Comedogenic rating: 0 Viscosity: medium Shelf life: 6–12 months Safety precautions: nontoxic, nonirritant when diluted.	Cold-pressed from the seeds	Rich in vitamins A, C, and E.	Restorative, regenerative, replenishing, anti-inflammatory, and antioxidant. Protects and repairs the skin from cellular damage, soothes irritation, and balances sebum production.

BOTANICAL INFORMATION	METHOD OF EXTRACTION	NUTRIENT CONTENT	THERAPEUTIC PROPERTIES
Jojoba oil *Simmondsia chinensis* Comedogenic rating: 2 Viscosity: medium Shelf life: 2–3 years Safety precautions: none	Cold-pressed from the seeds	Vitamins B and E and a range of minerals.	Nourishing, emollient, regenerative, toning, moisturizing, and anti-inflammatory. Prevents moisture loss and adds a protective barrier to the skin.
Olive oil *Olea europaea* Comedogenic rating: 2 Viscosity: light to medium Shelf life: 1–2 years Safety precautions: none	Cold-pressed and centrifuged	Vitamins A, E, and K and squalene.	Antioxidant, emollient, protective, soothing, and moisturizing. Repairs sun-damaged skin cells, fights damage by free radicals, and helps prevent premature aging.
Peach kernel oil *Prunus persica* Comedogenic rating: 2 Viscosity: light to medium Shelf life: 6–12 months Safety precautions: none	Cold-pressed and centrifuged	Vitamins A, E, B, and K.	Moisturizing, nourishing, protective, emollient, antiaging, and hydrating. Antioxidant, anti-inflammatory, and antimicrobial. Kind to dry and sensitive skin and promotes wound healing.
Pomegranate oil *Punica granatum* Comedogenic rating: 1 Viscosity: medium Shelf life: 1–2 years Safety precautions: none	Cold-pressed from the seeds	Vitamins A and E.	Antioxidant, anti-inflammatory, hydrating, regenerative, restorative, hydrating, and luxurious. Supports collagen production, protects against environmental and ultraviolet radiation damage and photoaging, improves the appearance of stretch marks and fine lines, and improves elasticity.
Pumpkin seed oil *Cucurbita pepo* Comedogenic rating: 2 Viscosity: light to medium Shelf life: 6–12 months Safety precautions: nontoxic, nonirritant when diluted.	Cold-pressed from the seeds	Rich in vitamins C and E.	Anti-inflammatory, moisturizing, wound healing, and rejuvenating. Protects against oxidation and cell damage, soothes irritation, and protects against free radicals.
Red raspberry seed oil *Rubus idaeus* Comedogenic rating: 0–1 Viscosity: medium Shelf life: 1–2 years Safety precautions: none	Cold-pressed from the seeds	Vitamin E, beta-carotene, and tocopherols.	Smoothing, emollient, anti-inflammatory, restorative, and hydrating. Prevents moisture loss and helps repair damaged skin. Has sun-protective properties.

BOTANICAL INFORMATION	METHOD OF EXTRACTION	NUTRIENT CONTENT	THERAPEUTIC PROPERTIES
Rice bran oil *Oryza sativa* Comedogenic rating: 2 Viscosity: light to medium Shelf life: 1–2 years Safety precautions: none	Cold-pressed	Rich in vitamin D and squalene.	Antioxidant, anti-inflammatory, soothing, and restorative. Protects against oxidation and cell damage, and protects against sun damage.
Rosehip seed oil *Rosa rubiginosa* Comedogenic scale: 1 Viscosity: thick Shelf life: 6–9 months refrigerated Safety precautions: none	Cold-pressed from the seeds	Rich in vitamins A, C, and E and beta-carotene.	Regenerative, nourishing, anti-aging, moisturizing, regenerative, and anti-inflammatory. Increases elastin content, promotes collagen formation, and improves the appearance of hyperpigmentation caused by sun damage and scar tissue.
Sea buckthorn oil *Hippophae rhamnoides* Comedogenic scale: 1 Viscosity: thick Shelf life: 1 year Safety precautions: none	Cold-pressed and CO_2 extracted from fruit and seed—it is a vibrant orange color	Rich in vitamins A, C, B_1, B_2, E, and K and beta-carotene.	Regenerative, wound healing, antioxidant, moisturizing, anti-inflammatory, and emollient. Prevents hyperpigmentation and age spots, is soothing to damaged skin, and is conditioning, especially to dry and mature skin.
Sesame oil *Sesamum indicum* Comedogenic scale: 2–3 Viscosity: light Shelf life: 1 year Safety precautions: none	Cold-pressed from the seeds	Rich in antioxidants and vitamin E.	Emollient, anti-inflammatory, wound healing, anti-aging, antifungal, and antibacterial. Regulates cell growth, offers a protective and balancing skin barrier, and is protective against sun damage.
Trauma oil Comedogenic scale: 2 Viscosity: light to medium Shelf life: 6–9 months Safety precautions: avoid exposure to the sun for 12 hours after application.	Cold-pressed from a blend of three herbal infusions: arnica, St. John's wort, and calendula	Rich in vitamin E.	Anti-inflammatory, softening, and regenerative. Promotes wound healing, alleviates pain, and relieves swelling.

Hydrosols

Hydrosols (also known as hydrolats, floral waters, and aromatic waters) are generally a by-product of the distillation process used to extract essential oils from plants.

When the water in a distillation still boils and steam rises through the plant matter, the aromatic oil is released. As the steam condenses, the oil separates and floats to the top, where it can be removed. The remaining water is called a hydrosol and contains all the beneficial essence of the plant, but in tiny amounts.

Hydrosols are brilliantly versatile. They make wonderful skin toners, cooling sprays, room or linen sprays, pillow sprays to help sleep, sprays to cool or soothe skin, or a more aromatic and therapeutic alternative to water in a skincare recipe. Because they are so gentle, they can safely be used for everyone, from tiny babies to pregnant and breastfeeding women and the elderly.

BOTANICAL INFORMATION	METHOD OF EXTRACTION	THERAPEUTIC PROPERTIES
Lavender hydrosol *Lavandula angustifolia* Shelf life: 1 year Safety precautions: none	Steam-distilled from the flowers	Anti-inflammatory, antioxidant, antimicrobial, antiviral, and acne-fighting. Soothing to rashes, allergies, sunburn, infections, and skin irritation.
Rosemary hydrosol *Rosmarinus officinalis* Shelf life: 1 year Safety precautions: none	Steam-distilled from the herb	Antibacterial, antiseptic, astringent, cleansing, decongestant, and energizing. Prevents damage by free radicals.

Herbs

Cheap and easily available (you can even grow your own!), herbs are a delicious way to care for your skin. Nothing beats the potent scent of fresh rosemary or thyme in your body scrub. It makes you feel as though you are showering in nature!

I use both dried and fresh herbs in these recipes, opting for fresh herbs if the recipe is for immediate or short-term use (such as the Cucumber and Rosemary Body Scrub on page 96), and dried when I want the product to last for weeks or months (as in the Rose Petal Vegan Bath Milk recipe on page 104).

BOTANICAL INFORMATION	THERAPEUTIC PROPERTIES
Lavender flowers *Lavandula angustifolia* Shelf life: 2 years when dried Safety precautions: none	Anti-inflammatory, antibacterial, antiseptic, wound healing, soothing to skin irritation, and sedative.
Nettle *Urtica dioica* Shelf life: 2 years when dried Safety precautions: none	Anti-inflammatory, antihistamine, antiseptic, firming, wound healing, and soothing to skin irritation.
Rose petals *Rosa damascena* Shelf life: 2 years when dried Safety precautions: none	Anti-inflammatory, antioxidant, antibacterial, and emollient. Supports the skin's immune system, soothing to skin irritation, and calming.
Rosemary *Rosmarinus officinalis* Shelf life: 2 years when dried Safety precautions: none	Anti-inflammatory, antioxidant, antibacterial, antiseptic, protective, and prevents free-radical damage.
Seaweed *Fucus vesiculosus* Shelf life: 2 years when dried Safety precautions: none	Contains magnesium and zinc. Anti-inflammatory, antibacterial, moisturizing, detoxifying, and soothing to skin irritation.
Thyme *Thymus vulgaris* Shelf life: 2 years when dried Safety precautions: none	Rich in vitamin C. Anti-inflammatory, antioxidant, antibacterial, antifungal, wound healing, and soothing to skin irritation.

Butters and Waxes

Butters and waxes, bursting with vitamins and fatty acids, are often hugely hydrating and deeply moisturizing. As they condition and soften our sometimes dry and damaged skin, they simultaneously deliver their rich nutrient load into the layers of the epidermis.

Butter recipes are best when whisked as the whisking aerates the butter, resulting in a soft and fluffy consistency which feels light and luxuriously delicious when applied to your skin!

Waxes create texture and structure in a balm or salve. However, they must always be used in conjunction with an oil or butter, otherwise the end product will be too hard in texture and impossible to apply to the skin. In sourcing waxes, you have two options: beeswax or a plant wax. As all the recipes in this book are vegan, they call for a beeswax substitute. I favor candelilla wax, derived from the plant of the same name. It is beautifully protective on the skin. Other beeswax alternatives are soy, jojoba, and rice bran wax, each with its own unique therapeutic benefits and aroma. All these are now more easily accessible because of the growing interest in plant-based products. One thing to look out for if you have been used to using beeswax: plant waxes have about twice the stiffening power of beeswax, so you will be using half the amount you have been accustomed to.

BOTANICAL INFORMATION	THERAPEUTIC PROPERTIES
Candelilla wax *Euphorbia antisyphilitica* Shelf life: 3–4 years Safety precautions: none	Supports the skin barrier, anti-aging, hydrating, and moisturizing. Note that it has twice the stiffening power of beeswax, so you will need much less.
Cocoa butter *Theobroma cacao* Shelf life: 3–4 years Safety precautions: none	Supports the skin barrier, anti-aging, hydrating, moisturizing, antioxidant, anti-inflammatory, emollient, and nourishing to damaged skin.
Shea butter *Butyrospermum parkii* Shelf life: 3–4 years Safety precautions: none	Replenishing, protective, moisturizing, and nourishing to dry and damaged skin.

Salts

Salts are an incredibly versatile weapon to have in your cosmetic arsenal, with their exfoliating, brightening, healing, firming, antiseptic, nourishing, anti-aging, and antibacterial properties, to name but a few!

Salts also pack a mighty mineral punch, containing calcium, sulfur, iodine, sodium, zinc, and potassium. The presence of magnesium, which binds to water and plumps up skin, also makes salts marvelously hydrating.

Salts are strongly anti-inflammatory, too. In the Muscle-soothing Bath Salts recipe on page 134, salts combine with the formidable anti-inflammatory power of CBD oil to create an awesome synergy, soothing muscular aches and pains while stimulating lymphatic circulation.

BOTANICAL INFORMATION	THERAPEUTIC PROPERTIES
Dead sea salts *Maris sal* Shelf life: indefinite when stored in a cool, dark place Safety precautions: none	Rich in minerals including magnesium, calcium, sulfur, iodine, sodium, zinc, and potassium. Anti-inflammatory. Exfoliates, stimulates circulation, increases lymphatic circulation helping to reduce excess fluid in the body, and reduces muscular aches, pain, and stiffness.
Epsom salts *Magnesium sulfate* Shelf life: indefinite when stored in a cool, dark place Safety precautions: none	Rich in magnesium and sulfur. Anti-inflammatory. Exfoliates and reduces muscular aches, pain, and stiffness.
Pink Himalayan salt *Sodium chloride* Shelf life: indefinite when stored in a cool, dark place Safety precautions: none	Rich in calcium, iron, zinc, chromium, magnesium, sulfate, and potassium. Anti-inflammatory, exfoliating, acne-fighting, antibacterial, detoxifying, and tonifying to the skin. Stimulates circulation and reduces muscular pain.
Magnesium flakes *Magnesium chloride* Shelf life: indefinite when stored in a cool, dark place Safety precautions: none	Anti-inflammatory, exfoliating, acne-fighting, and regenerative. Stimulates circulation, protective to the skin, repairs skin damage, increases lymphatic circulation which helps to reduce excess fluid in the body, and reduces muscular aches, pain, and stiffness.

Flours, Powders, and Clay

Flours and powders are some of the easier ingredients to use in skincare recipes. It only takes a little flour or powder to greatly enhance a recipe, and they are so easy to incorporate with other plant-based ingredients. They abound in goodness and are kind to our skin, bringing their own special nutrients to scrubs, masks, or soaks.

If you do not have a particular flour or powder to hand, don't worry! You can easily make your own by grinding the relevant material in a food processor or blender (for example, you can blitz oatmeal flakes to make oat flour).

Clay is earth's most primal element and has been part of our evolution as a species from the earliest times. It is a natural source of all the minerals used by humans for a variety of purposes. In terms of skincare, early humans are known to have used clay mixed with water to heal wounds, clean their skin, and soothe irritation—probably mimicking the animals who did this naturally. 21st-century humans continue this tradition—clay masks are today a multitasking beauty aid, acting as a powerful magnet for excess oil, impurities, trace metals, and other toxins, while simultaneously exfoliating and softening the skin.

BOTANICAL INFORMATION	THERAPEUTIC PROPERTIES
Blue spirulina powder *Arthrospira platensis* Shelf life: 2–3 years Safety precautions: none	Rich in vitamin E. Antioxidant, anti-inflammatory, hydrating, and firming.
Chickpea flour *Cicer arietinum* Shelf life: 6–12 months Safety precautions: none	Rich in zinc. Acne-fighting, antibacterial, wound healing, and brightening. Draws out impurities, exfoliates, evens skin tone, and improves skin immunity to bacteria.

BOTANICAL INFORMATION	THERAPEUTIC PROPERTIES
Green clay (montmorillonite or bentonite—the two are very similar) Shelf life: indefinite when stored in a cool, dark place Safety precautions: none	Antioxidant, anti-inflammatory, wound healing, cleansing, detoxifying, and decongesting. Toning to the skin and revitalizing to the complexion. Exfoliates and draws out impurities.
Kale powder *Brassica oleracea acephala* Shelf life: 2 years Safety precautions: none	Rich in vitamins K, A, and C. Antioxidant, regenerative, promotes collagen production, exfoliates, detoxifies, and prevents damage by free radicals.
Matcha powder *Camellia sinensis* Shelf life: 1 year Safety precautions: none	Antioxidant, anti-inflammatory, rejuvenating, brightening, cleansing, detoxifying, and regenerative. Evens skin tone, balances skin's own oil content, and increases skin elasticity. Combats damage by free radicals and oxidative stress.
Mustard seed powder *Brassica alba* Shelf life: 3–4 years Safety precautions: none	Antioxidant, anti-inflammatory, wound healing, cleansing, warming, and pain relieving. Stimulates circulation.
Oat flour *Avena sativa* Shelf life: 1–2 years Safety precautions: none	Antioxidant, anti-inflammatory, exfoliating, wound healing, cleansing, and moisturizing. Draws out impurities.
Organic oats *Avena sativa* Shelf life: 1–2 years Safety precautions: none	Antioxidant, anti-inflammatory, exfoliating, wound healing, cleansing, and moisturizing. Draws out impurities.
Wheatgrass powder *Triticum aestivum* Shelf life: 2–3 years Safety precautions: none	Rich in vitamins and amino acids. Anti-inflammatory, regenerative, anti-aging, detoxifying, and cleansing.
White clay (kaolin) Shelf life: indefinite when stored in a cool, dark place Safety precautions: none	Antioxidant, anti-inflammatory, exfoliating, and cleansing. Draws out impurities, reduces swelling in the skin, and stimulates circulation.

Fruit and Vegetables

There is something so satisfying about using fresh fruit and vegetables in masks and scrubs. Not only do they smell divine, but they are seriously effective in treating various skin conditions, especially sensitive or lackluster skin. The symbiosis of minerals, nutrients, enzymes, and vitamins in our fresh produce nourishes the skin; the different textures they offer allow us to create splendid exfoliators while also soothing irritation. Couple blended fruit and vegetables with therapeutic essential and carrier oils, and you have a skincare dream team!

BOTANICAL INFORMATION	THERAPEUTIC PROPERTIES
Avocado *Persea americana* Shelf life: 3–5 days Safety precautions: none	Rich in vitamins A, B, and E, and in omega-9. Anti-aging, anti-inflammatory, protective, regenerative, and conditioning. Increases collagen production, protects against UV damage, repairs damaged skin, and kind to sensitive and damaged skin.
Banana *Musa acuminata* Shelf life: 1 week Safety precautions: none	Rich in vitamin A. Moisturizing, hydrating, and restorative. Helps increase collagen production and softens skin texture.
Blueberry *Vaccinium corymbosum* Shelf life: 1–2 weeks Safety precautions: none	Rich in vitamin E. Anti-aging, protective, and regenerative. Smoothes fine lines, increases elasticity of the skin, and neutralizes damage from free radicals.
Cucumber *Cucumis sativus* Shelf life: 2 weeks Safety precautions: none	Anti-inflammatory, moisturizing, and regenerative. Rebuilds damaged skin and reduces swelling and puffiness.
Mango *Mangifera indica* Shelf life: 2 weeks Safety precautions: none	Antioxidant, anti-inflammatory, hydrating, and revitalizing. Soothes skin irritation.
Raspberry *Rubus idaeus* Shelf life: 5 days Safety precautions: none	Rich in vitamin C. Antioxidant, anti-aging, anti-inflammatory, and moisturizing. Evens skin tone and protects from UV damage.

Other Ingredients

Some ingredients, such as castile soap and maple syrup, have been chosen for their function. Others, such as hempseeds, are there for their texture. All bring their own therapeutic properties, which are listed in the chart below.

BOTANICAL INFORMATION	METHOD OF EXTRACTION	THERAPEUTIC PROPERTIES
Agave nectar *Agave americana* and *Agave tequilana* Consistency: thick Shelf life: 2–3 years Safety precautions: none	Sap is extracted from the core of the plant, filtered, and then heated to convert the carbohydrates into sugars.	Moisturizing, softening, and brightening. Calms irritated skin.
Aloe vera gel *Aloe barbadensis* "Miller" Consistency: gel-like Shelf life: 2–3 years if refrigerated once opened Safety precautions: none	Aloe vera gel is stored in the leaf of the plant. It is easily extracted using a knife.	Good skin penetration properties. Reduces scarring, moisturizes, hydrates, rejuvenates, and assists wound healing. Soothing for sun damage and skin irritation.
Castile soap *Sapo hispaniensis* Consistency: thin Shelf life: 3 years Safety precautions: do not ingest	Originally made exclusively from olive oil, using the classic saponification process. Other vegetable oils now increasingly used.	A gentle cleanser which will effectively remove dirt while leaving the skin soft.
Coconut milk *Cocos nucifera* Consistency: liquid Shelf life: 7–10 days Safety precautions: none	Extracted from grated coconut meat after pressing or squeezing, with or without the addition of warm water.	Moisturizing, especially to irritated skin.
Coffee *Coffea arabica* Consistency: granular Shelf life: 1 year Safety precautions: none	Ground coffee beans.	Anti-inflammatory and antioxidant. Rejuvenates and brightens the complexion.
Glycerine Consistency: thick Shelf life: 2–3 years Safety precautions: none	Process of distillation and saponification.	Efficient humectant which helps retain moisture in the skin's upper epidermal layers.

BOTANICAL INFORMATION	METHOD OF EXTRACTION	THERAPEUTIC PROPERTIES
Hempseeds *Cannabis sativa* Consistency: seeds Shelf life: 6 months Safety precautions: none	Dried seeds collected from harvested flowers.	Anti-inflammatory and rejuvenating.
Maple syrup *Acer saccharum* Consistency: thick Shelf life: 2–3 years Safety precautions: none	Holes are drilled into the maple tree trunk and the sap is collected and heated, leaving the concentrated syrup.	Antioxidant and can help fight wrinkles, dryness, redness, and inflammation.
Organic brown sugar (sucrose) Consistency: granular Shelf life: indefinite Safety precautions: none	Shredding and squeezing sugar canes.	Anti-bacterial and exfoliating.
Soy yogurt *Soja hispida* Consistency: viscous Shelf life: 7–10 days Safety precautions: do not use if suffering from a soy allergy	Process of fermentation.	Moisturizing. Soothes skin irritation and helps to prevent breakouts.
Vanilla extract *Vanilla planifolia* Consistency: liquid Shelf life: indefinite Safety precautions: nontoxic, nonirritant when diluted.	Washing and soaking ground vanilla beans in a solution of water and alcohol.	Antioxidant, anti-inflammatory, antimicrobial, and aphrodisiac. Neutralizes free radicals, helping to slow the signs of aging.
Witch hazel *Hamamelis virginiana* Consistency: thin Shelf life: 1–2 years Safety precautions: none	Steam-distilled from the dried leaves and bark.	Anti-inflammatory, astringent, antimicrobial, and wound healing.

CHAPTER 4

Your Blending Kitchen

You won't need any specialized equipment for the recipes in this book. While it is likely that you already have most of what you need in your kitchen, I do recommend that you keep a separate set of bowls and beakers for your beauty recipes. Many of the ingredients, especially the CBD oil, the essential oils, and the botanical ingredients, will leave a taste on your equipment which you will not want to impart to food.

Useful Equipment

Here is what you will need in order to become a master of your CBD beauty recipes.

BLENDER

A food blender or a Nutribullet™ machine with a large jug will do the job perfectly, whether making the butter-based recipes and creams, or grinding hempseeds, oats, or other dry ingredients.

HAND MIXER WITH WHISK ATTACHMENT

You can use a tabletop mixer or a hand mixer with a whisk attachment for the recipes which require whipping. I tend to use the hand mixer unless I am making larger batches of a recipe. For these, a food processor is ideal. It is best to use a stainless-steel or glass bowl with your mixer (see below).

If you do not have an electric mixer, you can use a hand whisk, although it will be a little more work! A small hand whisk is useful for small amounts, such as for the Gentle Body Cleanser on page 94.

DOUBLE BOILER

A double boiler is comprised of two pans, one on top of the other, and is used to melt butters and waxes. The material to be melted goes in the top pan; the second pan contains boiling water. Gently stir the material until melted, then remove the double boiler from the heat.

DIGITAL KITCHEN SCALE

Some of the ingredient amounts are small, so use a digital scale to ensure your measurements are accurate.

MEASURING SPOONS

Stainless-steel tablespoons and teaspoons will be super-helpful, as stainless steel is quick to clean and easily sterilized, but you can also use traditional food measuring spoons.

GLASS BOWL AND BEAKER

Glass beakers will give you an easy pour for the oil-based recipes, and glass containers are easy to clean and sterilize.

Above: **An electric hand whisk makes light work of whipping butters to the desired consistency.**

Left: **Keep your beauty recipe equipment separate from your everyday cooking utensils.**

STAINLESS-STEEL BOWL AND BEAKER

I like to use stainless steel for my main blending equipment. It is ideal for melting butters and waxes, as well as gentle reheating if ingredients begin to resolidify.

STIRRER

You will need a stirrer for many of the recipes. Glass is easier to clean, but a wooden stirrer will also be fine.

SPATULA

A good-quality spatula will ensure that you don't waste any of your mixture as you transfer it to a storage container. I favor a silicone-topped spatula as the most effective at getting the last traces of mixture out.

SILICONE MOLDS

You will need molds for the Lavender Sleep Massage Bar (see page 110), and silicone molds are easy to peel. You can use paper muffin cases instead, but make sure that your bars are fully set before peeling off the paper.

KITCHEN GRATER

You will need a standard grater to grate the cocoa butter for the Rose Petal Vegan Bath Milk (see page 104).

Storing Your Ingredients

Efficient storage will keep your ingredients stable and fresh and will prolong their shelf life.

I store dry ingredients like oat flour and hemp seeds in mason jars, in a dark cupboard. I keep vegetable and essential oils and hydrosols in dark glass bottles or jars (amber, navy, or dark green), storing them in a cool, dry cupboard and making sure to seal them tightly after each use. My hydrosols are happy chilling out in the fridge.

For storing my completed products, as opposed to my ingredients, I favor glass containers. They are very versatile and come in a range of colors, shapes, and sizes. In each recipe I suggest the ideal storage container, but it is useful to have on hand a selection of 1-oz (30-ml) and 2-oz (60-ml) bottles with a variety of closures, including screw top, flip cap, and pumps, as well as jars with screw lids. Again, I prefer dark glass in order to protect the CBD oil and the natural ingredients.

It is a great idea to label ingredients and products with the date you made or opened them (as appropriate), so you can track their freshness.

Shelf life of products

Every recipe in this book offers advice on how long the product will stay fresh and how to store it. Some recipes are for one to three uses and others will keep for months (or even years in the case of salts). Generally speaking, a recipe containing fresh vegetable or fruit ingredients will need to be used within a few days. The secret for using these perishable products is the fresher, the better.

A good rule of thumb is that if the recipe is oil-based, it will last longer than a water-based recipe. Microbes and bacteria love moisture, whereas oil-based recipes do not offer a welcoming environment in which they can flourish. In case you are concerned about our water-based recipes, let me reassure you that they always include natural antioxidants to keep the baddies away!

FACIAL AND BODY OILS

Oils are typically not susceptible to mold caused by microbial and bacterial activity. However, over time they will oxidize and eventually turn rancid. They will all keep fresh for two months if you keep them cool and away from direct heat and sunlight. I always store oils in dark bottles to protect them, from the day they were created. If you want to make bigger batches and extend your product's shelf life, add 1% of vitamin E to your recipe. Vitamin E is a natural antioxidant which will bring an additional layer of protection to your recipe (see page 27).

BALMS, SALVES, AND BUTTERS

Like oil-based products, balms, salves, and butter-based products have no water content and therefore constitute a hostile environment for microbes and bacteria. However, the carrier oils will tend to go rancid after a year. If you want to extend their shelf life, feel free to add 1% of vitamin E to your recipe for an antioxidant boost.

SALTS

Recipes containing salts are very stable and will keep for years if stored in a dry, airtight container.

HYDROSOLS

Recipes which contain floral waters will last between a week and a month, depending on the natural antioxidant strength of the other ingredients in the recipe. They are best stored in the refrigerator.

Right: **The Tea Tree Hand Sanitizing Gel (see page 114) can be kept in a small dark glass bottle.**

CHAPTER 5

Facial Love

Less is more with natural facial products. Look at the list of contents on the packaging of any popular chemical-based cosmetic. There will probably be anything up to 50 ingredients, many with long scientific names. What you see in the short recipes below is exactly what you get—no lurking nasties such as parabens, industrial solvents, sodium lauryl sulfate, formaldehyde, even phthalates (a key component in plastics). It has been found that up to 60% of chemicals commonly used in cosmetics are absorbed rapidly into the skin and then, via the bloodstream, into the body's major organs. Instead, you could choose to let your skin absorb pure, effective, and beautiful alternatives provided by Mother Nature. I feature my favorites in this book. They all include the amazing CBD oil which has become a staple in my natural skincare formulation kit and will remain vitally important as I continue to curate skincare recipes.

Coconut and Pomegranate Whipped Makeup Remover

A basic principle of chemistry is that "oil dissolves oil." Using an oil-based cleanser rather than a soap or foam cleanser can be super-effective at removing debris and grime from your pores. Elegant yet effective, this coconut, pomegranate, and CBD oil makeup remover will lift makeup and dirt from your skin to leave it thoroughly cleansed and deliciously soft to the touch.

4 tablespoons coconut oil

½ teaspoon pomegranate oil

½ teaspoon CBD oil

Equipment

Hand mixer with whisk attachment, or a hand whisk

Spatula

Stainless-steel bowl

2-oz (55-g) glass jar with airtight lid

Preparation time: 10 minutes

Makes 2 oz (55 g)

1 Whip the coconut oil with the hand mixer on high for three minutes, stopping occasionally to scrape the oil from the sides to the center of the bowl using a spatula.

2 Turn off the mixer and add the pomegranate and CBD oils. Whip for a further minute to thoroughly combine the oils. When the mixture is fluffy in texture, you will know it is ready.

3 Transfer the mixture to the glass jar using the spatula, and seal.

STORAGE

This product will last for two months if you store it in an airtight container and keep it away from direct sunlight and heat. If your makeup remover becomes slightly hard over time, don't panic! Just warm it between your fingertips before application.

TO USE

Take a generous amount of the product between your fingertips. Massage over your face and neck. Remove gently with a soft, damp cloth.

APOTHECARY NOTE

You can make this recipe without the pomegranate oil. The coconut and CBD oil combined will work perfectly. However, it is the pomegranate oil which adds a luxurious after-feeling on your skin, so I recommend going for that little touch of opulence!

Baobab Cleansing Oil

Castor oil is more viscous than many other vegetable oils but, despite its thick consistency, it has a very low comedogenic score—this means that it will not clog your pores. Baobab oil is a stunning vegetable oil of African origin, rich in vitamins A, D, and E, with beautiful moisturizing and tonifying properties. The combination of ingredients below will leave your skin squeaky clean without stripping any of its natural oils.

2 tablespoons baobab oil

1 tablespoon sesame oil

1 teaspoon castor oil

1 teaspoon glycerine

1 teaspoon CBD oil

3 drops grapefruit essential oil

1 drop ylang ylang essential oil

Equipment

Glass beaker or stainless-steel bowl

Glass or wooden stirrer

2-fl oz (60-ml) glass bottle with dropper cap or flip lid

Spatula

Preparation time: 10 minutes

Makes 2 fl oz (60 ml)

1 Combine the baobab oil, sesame oil, castor oil, glycerine, and CBD oil in the bowl or beaker.

2 Add the grapefruit and ylang ylang oils and stir gently.

3 Transfer the mixture to the glass bottle using a spatula, and seal.

STORAGE

As there is no water in this cleansing oil, it will keep for up to three months if stored away from direct sunlight and heat.

TO USE

Shake before use. Massage the oil into your skin, using upward circular motions to loosen makeup and dirt, and avoiding the eye area. When your entire face is cleansed, use cotton pads or a warm, damp cloth to rinse your skin. Follow with a toner (such as the Lavender and Rosemary Toner on page 68) or hydrosol (see page 46).

APOTHECARY NOTE

If you want to use this cleansing oil to remove eye makeup, leave out the grapefruit and ylang ylang oils as they can be irritating to the sensitive eye area.

Lavender and Rosemary Toner

A toner was once the PH balancing middle step in a "cleanse, tone, and moisturize" skincare routine. However, this traditional three-step ritual has not been in fashion for a long time. Now, in the world of botanical skincare, simple yet elegant hydrosols (see page 46) are bringing toners back into fashion. Spritzing your skin with a floral water after cleansing and just before applying a vitamin-rich moisturizer or serum is refreshing and soothing to your skin. Astringent rosemary and witch hazel make this recipe ideal for oily skin (but also suitable for all skin types), while the CBD oil and lavender offer soothing and anti-inflammatory support for your skin.

2 tablespoons lavender hydrosol

2 teaspoons rosemary hydrosol

2 teaspoons witch hazel

1 teaspoon glycerine

1 teaspoon CBD oil

Equipment

Glass or stainless-steel pitcher (jug)

Glass or wooden stirrer

2-fl oz (60-ml) glass bottle with pump dispenser

Preparation time: 5 minutes

Makes 2 fl oz (60 ml)

1 Combine the lavender hydrosol, rosemary hydrosol, witch hazel, glycerine, and CBD oil in the pitcher (jug). Stir gently.

2 Transfer the mixture to the glass bottle and seal.

STORAGE

Your lavender and rosemary toner will keep fresh for up to three months if stored away from direct sunlight and heat.

TO USE

Shake before use. After cleansing, either spritz your skin with the toner or apply with a cotton pad. Pat into the skin.

APOTHECARY NOTE

As we explored in the introduction, hydrosols are a by-product of the steam distillation process which is used to extract essential oil from plant material. These aromatic or floral waters are therapeutic in their own right, as well as being a beautiful and advantageous addition to your beauty kit.

Antioxidant Matcha Facial Scrub

The green superfood matcha has been made famous during the "matcha latte revolution" in the last few years, and has now gained a firm foothold in the beauty world. No wonder, as matcha is known as an anti-aging skin hero owing to its ability to stimulate elastin production, while also containing powerful antioxidant, anti-aging, anti-inflammatory, and exfoliant properties. Combined with the potent anti-inflammatory properties of CBD oil, this is a powerhouse of a facial scrub!

2 teaspoons organic oats

1 teaspoon matcha powder

1 teaspoon white clay

½ teaspoon castor oil

½ teaspoon rice bran oil

½ teaspoon maple syrup

½ teaspoon CBD oil

2 drops carrot seed essential oil

2 drops ylang ylang essential oil

2 drops frankincense essential oil

Equipment

Glass or stainless-steel bowl

Glass or stainless-steel beaker

Glass or wooden stirrer

1-oz (30-g) glass jar with airtight lid

Preparation time: 10 minutes

Makes 1 oz (30 g), enough for three uses

1 Combine all the dry ingredients—organic oats, matcha powder, and white clay—in the bowl.

2 Combine the castor oil, rice bran oil, maple syrup, CBD oil, and all three essential oils in the beaker.

3 Slowly stir the blend of oils into the combined dry ingredients. If the mixture is too thin for your liking, add a few more oats. If it is too thick, add a little more rice bran oil until you get the desired texture.

4 Transfer the mixture to the glass jar and seal.

STORAGE

This recipe has a short shelf life and should be used within a week. Store in a cool place, ideally a refrigerator.

TO USE

Dampen your face with warm water. Gently massage a small amount of the scrub into your skin for 1–2 minutes, using circular motions. If the scrub feels harsh on your skin, gradually add a little water. Rinse your face with warm water and wipe away any scrub residue with a warm, damp cloth until your face is clean.

APOTHECARY NOTE

The 12th-century Japanese Zen-Buddhist monk Myōan Eisai discovered green tea on a visit to China. Finding that drinking this tea improved his Zen meditation and created a state of calm alertness, he brought it back with him to Japan—where it quickly became the basis of the sophisticated art of the Japanese tea ceremony. Matcha powder comes from green tea, which in turn is derived from the leaves and buds of *Camellia sinensis*, an evergreen shrub with small white flowers.

Purifying Hemp Facial Scrub

This facial scrub is brilliant if you are dealing with clogged pores or lackluster skin. The combination of wheatgrass, kale, and blue spirulina powders work in sweet synergy to infuse your skin with purifying nutrients. The ground hempseeds not only deliver effective physical exfoliation, but also help to regulate oily skin and clean clogged pores—leaving a noticeably brighter complexion.

2 teaspoons chickpea flour

1 teaspoon ground hempseeds

½ teaspoon kale powder

¼ teaspoon wheatgrass powder

¼ teaspoon blue spirulina powder

½ teaspoon CBD oil

2 teaspoons hempseed oil

2 drops marjoram essential oil

2 drops rosemary essential oil

2 drops peppermint essential oil

Equipment

Glass or stainless-steel bowl

Glass or stainless-steel beaker

Glass or wooden stirrer

2-oz (55-g) glass jar with lid

Preparation time: 10 minutes

Makes 2 oz (55 g), enough for three uses

1 Combine all the dry ingredients—chickpea flour, ground hempseeds, kale powder, wheatgrass powder, and blue spirulina powder—in the bowl.

2 Combine the CBD oil, hempseed oil, and the three essential oils in the beaker.

3 Slowly stir the blended oils into the combined dry ingredients. If the mixture is too thin for your liking, add a little more chickpea flour. If it is too thick, add a little more hempseed oil until you get the desired texture.

4 Transfer the mixture to the glass jar and seal.

STORAGE

This recipe has a short shelf life and should be used within a week. Store in the refrigerator.

TO USE

Dampen your face with warm water. Gently massage a small amount of the scrub into your skin in circular motions for 1–2 minutes. If the scrub feels harsh on your skin, gradually add a little water. Rinse your face with warm water and wipe off any scrub residue with a warm, damp cloth until your face is clean.

APOTHECARY NOTE

If you cannot source ground hempseeds, you can grind whole hempseeds with a pestle and mortar, or blitz in a food processor. Be warned, they grind very quickly, so only blitz for a few seconds!

Probiotic Berry Mask

One of the main contributors to aging and stressed skin is the presence of free radicals (see page 41). The good news is that the cellular damage they cause can be fought using natural and easily available products. This probiotic face mask harnesses the powerful antioxidant properties of raspberries and blueberries, known to delay the process of cell oxidation. Use it weekly to nourish your skin and to boost its own defenses.

1 heaped tablespoon blueberries

1 heaped tablespoon raspberries

½ banana

1 heaped tablespoon soy yogurt

½ teaspoon CBD oil

1 drop grapefruit essential oil

Equipment

Standard kitchen blender

Glass or stainless-steel bowl

Glass or wooden stirrer

Preparation time: 5 minutes

Makes 1 oz (30 g) for single immediate use

1 Put the blueberries, raspberries, and banana into the blender and blitz for 30–60 seconds, until the mixture is liquefied.

2 Transfer the mixture to the bowl and stir in the yogurt, CBD oil, and grapefruit essential oil.

TO USE

Apply the mask to your face immediately and let it work its magic for 15 minutes. Rinse your face with warm water and pat your skin dry with a soft cotton cloth.

APOTHECARY NOTE

The soy yogurt in this recipe has a secret superhero effect on your skin! Not only does soy yogurt contain probiotics, but it also contains vegan-friendly lactic acid. It is known as a gentle and effective exfoliator and for its anti-aging effects.

Anti-inflammatory Mango Mask

This face mask is quick and easy to make and use. CBD oil is a powerful anti-inflammatory ingredient which helps promote skin health, and is partnered here with the nutrient-dense mango fruit. Rich in one of nature's most potent antioxidants, vitamin C, the mango also has good levels of vitamins A, B$_6$, E, and K, along with valuable mineral content. Combined in this recipe, CBD oil and mango provide powerful skin support and are a formidable ally in the battle against skin cell damage.

½ mango (not too ripe, or the mixture will be too thin)

1 teaspoon agave nectar

1 teaspoon CBD oil

1 drop Roman chamomile essential oil

1 drop lavender essential oil

Equipment

Standard kitchen blender

Glass or stainless-steel bowl

Glass or wooden stirrer

Preparation time: 5 minutes

Makes 1 oz (30 g) for single immediate use

1 Put the mango in the blender and blitz for 30 seconds on a high setting, until the fruit is liquefied.

2 Transfer to the bowl and stir in the agave nectar, CBD oil, and the Roman chamomile and lavender essential oils.

TO USE

Apply the mask to your face immediately and let it work its magic for 15 minutes. Rinse your face with warm water and pat your skin dry with a soft cotton cloth.

French Green Clay Mask

Elegant French women have been using green clay for centuries as their purifying go-to beauty ingredient—and who is going to argue with the French when it comes to beauty? Soft green clay, known as montmorillonite, has been mined for centuries in western France. It is now widely sought after in the beauty and hair care sectors. Mineral-rich and highly absorbent, this clay is excellent at absorbing dirt, debris, impurities, oil, and toxins from the skin. It is particularly effective for oily skin and hair. Green clay is also known for stimulating circulation and having a toning and firming effect. These properties are perfectly balanced in this recipe by the soothing and calming combination of avocado and a lavender hydrosol.

½ ripe avocado (flesh only)

¼ cucumber

2 teaspoons lavender hydrosol

1 teaspoon agave nectar

1 teaspoon aloe vera gel

1 teaspoon matcha powder

1 teaspoon green clay

½ teaspoon CBD oil

1 drop geranium essential oil

1 drop cedarwood essential oil

Equipment

Standard kitchen blender

Glass or stainless-steel bowl

Glass or wooden stirrer

1-oz (30-g) glass jar and lid

Preparation time: 5 minutes

Makes 1 oz (30 g), enough for two uses

1 Blitz the avocado and cucumber in the blender for 1 minute until the mixture is liquefied.

2 Add the lavender hydrosol, agave nectar, aloe vera gel, matcha powder, green clay, and CBD oil, then blitz for a further 30 seconds.

3 Stir in the geranium and cedarwood essential oils.

4 Transfer the mixture to the glass jar and seal.

STORAGE
Store in the refrigerator for no longer than one week.

TO USE
Apply the green clay mask to your face and let it work for 15 minutes. Wipe off with a damp cloth and rinse with warm water. The clay can be stubborn, so be sure to remove it completely from your face! Pat your skin dry with a soft cotton cloth.

APOTHECARY NOTE
Unlike traditional clay masks, this one will not harden (because of the nectar and hydrosol ingredients).

Aging Gracefully Skin Elegance Serum

Rosehip seed oil and carrot seed essential oil are both rich in vitamin A and beta-carotene, which makes them an impressive duo for this powerful skin serum. They are known to slow the signs of aging on your skin and repair skin damage. In addition, they penetrate the skin rapidly and deeply, for maximum impact.

2 tablespoons jojoba oil

2 teaspoons rice bran oil

1 teaspoon rosehip seed oil

1 teaspoon borage oil

1 teaspoon sea buckthorn oil

1 teaspoon CBD oil

4 drops carrot seed essential oil

3 drops frankincense essential oil

1 drop Palo Santo essential oil

Equipment

Glass or stainless-steel beaker

Glass or wooden stirrer

2-fl oz (60-ml) glass bottle with dropper cap or flip lid

Preparation time: 5 minutes

Makes 2 fl oz (60 ml), enough for around 30 nightly uses

1 Combine the jojoba, rice bran, rosehip seed, borage, sea buckthorn, and CBD oils in the beaker.

2 Gently stir in the carrot seed, frankincense, and Palo Santo essential oils.

3 Transfer the mixture to the glass bottle and seal.

STORAGE

As there is no water in this serum, it will keep for up to three months if stored away from direct sunlight and heat.

TO USE

Massage into thoroughly cleansed skin with upward circular strokes before going to bed at night.

APOTHECARY NOTE

During the night, our skin behaves differently from during the day. In the daytime, the skin is busy fighting environmental pollutants and damage from ultraviolet radiation. At night, it is not only at rest, but it also naturally kicks into repair mode. This mode is facilitated by increased nocturnal production of two super-hormones, melatonin and human growth hormone. These accelerate the skin's natural regenerative ability, as well as increasing its production of antioxidants. Therefore, a facial serum rich in skin-repairing plants is most effective when applied at night.

Red Raspberry Seed Facial Oil

Red raspberries were created by the fruit gods to heal and soothe our stressed skin! This oil, extracted from the seeds, is packed with essential fatty acids and omega-3, which help reduce inflammation, while the oil's beta-carotene levels give it strong antioxidant properties. This is the oil that keeps on giving—improving elasticity, smoothing wrinkles and sagging skin, helping prevent trans-epidermal water loss, and deeply hydrating tired skin. Working synergistically with the anti-inflammatory and antioxidant properties of CBD oil, this facial oil will quickly become your go-to product for skin in need of TLC.

2 tablespoons apricot oil

2 teaspoons red raspberry seed oil

1 teaspoon hempseed oil

1 teaspoon pumpkin seed oil

1 teaspoon CBD oil

1 teaspoon peach kernel oil

3 drops lavender essential oil

3 drops frankincense essential oil

3 drops ylang ylang essential oil

2 drops Roman chamomile essential oil

Equipment

Glass or stainless-steel beaker

Glass or wooden stirrer

2-fl oz (60-ml) glass bottle with dropper cap or flip lid

Preparation time: 5 minutes

Makes 2 fl oz (60 ml)

1 Combine the apricot, red raspberry seed, hempseed, pumpkin seed, CBD, and peach kernel oils in the beaker.

2 Gently stir in the lavender, frankincense, ylang ylang, and Roman chamomile essential oils.

3 Transfer the mixture to the glass bottle and seal.

STORAGE

As there is no water in the ingredients, this mixture will keep for up to three months if stored away from direct sunlight and heat.

TO USE

Massage into cleansed skin, using upward circular strokes. This oil is suitable for morning or evening application, but I prefer to apply it in the morning because it is so light and fresh. It also absorbs really well, so it tends to leave your skin treated and conditioned but not greasy or shiny.

Shea and Apricot Facial Cream

Shea butter is a skin superfood, abounding in essential fatty acids and rich in vitamins A, C, and E. An outstanding moisturizer, it drenches the skin with nutrients, soothes and repairs stressed and damaged skin, and helps boost collagen production and skin elasticity.

½ oz (15 g) shea butter

1 teaspoon CBD oil

1 teaspoon apricot oil

1 teaspoon rice bran oil

1 drop carrot seed essential oil

1 drop ylang ylang essential oil

1 drop petitgrain essential oil

Equipment

Hand mixer with whisk attachment, or a hand whisk

Stainless-steel bowl

Spatula

1-oz (30-g) glass jar with airtight lid

Preparation time: 5 minutes

Makes 1 oz (30 g)

1 Whip the shea butter with the hand mixer on high for three minutes. You will start to see the shea butter become creamier.

2 After 3 minutes, start gradually adding the CBD, apricot, and rice bran oils, along with the three essential oils.

3 Whip until all the ingredients are combined and the mixture is creamy in texture.

4 Transfer the mixture to the glass jar using a spatula, and seal.

STORAGE

This product will last for two months if you store it in an airtight container, away from direct sunlight and heat.

TO USE

Apply generously to thoroughly cleansed skin, using upward circular movements.

APOTHECARY NOTE

This cream is wonderful all year round. It is particularly supportive during winter, when our skin is exposed to harsh weather. It is also immensely soothing and comforting to skin which has been overexposed to the sun!

Rosehip Eye Serum

Rosehip seed oil and CBD oil are a perfect pairing for skincare. Both oils have robust anti-inflammatory properties that are particularly useful in treating stressed, dehydrated, acne-prone, or irritated skin. Rosehip seed oil beautifully complements CBD because it abounds in beauty essentials. Its essential fatty acids help rebuild the skin's own protective barrier against environmental stresses such as air pollutants and UV radiation. The omega-3 content helps address hyperpigmentation and the appearance of fine lines and aging skin, so it's the perfect ingredient as the basis for an eye serum. All in all, this serum is a winning combination.

1½ teaspoons rosehip seed oil

½ teaspoon apricot oil

½ teaspoon rice bran oil

½ teaspoon CBD oil

Equipment

Glass or stainless-steel beaker

Glass or wooden stirrer

½-fl oz (15-ml) glass bottle with lid

Preparation time: 5 minutes

Makes ½ fl oz (15 ml)

1 Combine the rosehip seed, apricot, rice bran, and CBD oils.

2 Transfer to the glass bottle and seal.

STORAGE

This serum will keep for up to three months if stored away from direct light and heat. I keep mine in the refrigerator because the cooler the product is, the greater the effect on puffy eyes and irritation around the eye area.

TO USE

After thoroughly cleansing your face, apply the serum using circular strokes around the eye area.

Aloe Blemish Gel

We have all had those untimely breakouts or blemishes which require speedy intervention to halt the inflammatory reaction which is causing the problem. Tea tree essential oil and CBD oil possess powerful antimicrobial and antibacterial properties, while witch hazel gets to work drying up excess oil in the pores. Aloe vera, marjoram, and lavender oils, known for their potent anti-inflammatory and soothing properties, complete this rapid-response team.

2 teaspoons aloe vera gel

1 teaspoon witch hazel

¼ teaspoon CBD oil

10 drops lavender essential oil

6 drops marjoram essential oil

4 drops frankincense essential oil

4 drops tea tree essential oil

Equipment

Glass or stainless-steel beaker

Glass or wooden stirrer

½-fl oz (15-ml) glass bottle with pump dispenser

Preparation time: 5 minutes

Makes ½ fl oz (15 ml)

1 Combine the aloe vera gel, witch hazel, and CBD oil in the beaker.

2 Gently stir in the drops of lavender, marjoram, frankincense, and tea tree essential oils.

3 Transfer the mixture to the glass bottle and seal.

STORAGE

This recipe contains no preservative, so, despite the antibacterial properties of the tea tree oil, it is recommended that you use the mixture within a month.

TO USE

Thoroughly cleanse your skin, paying special attention to the affected area. Using a cotton bud, apply the gel to the affected area and allow it to absorb into the skin fully. Repeat this as often as required, but at least twice a day.

Argan Lip Balm

Argan oil is a light, non-greasy oil, which is quickly absorbed by the skin. Like CBD oil, it has substantial antioxidant and anti-inflammatory properties, owing to its excellent range of fatty acids. This balm is packed with skin-soothing nutrients. I always keep a little jar of this liquid gold in my purse or close at hand for whenever my lips are in need of a little TLC.

2 teaspoons cocoa butter

2 teaspoons candelilla wax

2 teaspoons coconut oil

½ teaspoon argan oil

½ teaspoon CBD oil

5 drops peppermint essential oil

5 drops grapefruit essential oil

5 drops orange essential oil

Equipment

Double boiler

Stainless-steel bowl

Glass stirrer

6 small (0.16-fl oz/5-ml) glass pots with lids, or a 1-fl oz (30-ml) glass jar with lid

Preparation time: 20 minutes

Makes 1 oz (30 g)

1 Melt the cocoa butter, candelilla wax, and coconut oil in the double boiler.

2 Keeping the bowl over the heat, stir in the argan and CBD oils.

3 Remove from the heat and gently stir in the peppermint, grapefruit, and orange essential oils.

4 Transfer to the six small glass jars (or single larger jar) and seal.

STORAGE

The lip balms will keep for up to 12 months.

TO USE

Apply to your lips as often as required.

APOTHECARY NOTE

Vegan skincare has become so popular in the last few years! Even if you are not committed to a vegan lifestyle, you could take your first step by converting to vegan skincare. This is relatively easy until you need to use a wax to make a balm or salve. A vegan alternative to beeswax is candelilla wax, extracted from the leaves of a desert plant found in northern Mexico and the south-west of the United States. It was originally used for candle making, but its ability to prevent moisture loss and protect our skin has made it very popular for skincare preparations. It is similar to beeswax in terms of texture and melting points. However, if you are substituting candelilla in a recipe calling for beeswax, you need to use half the amount, as candelilla has twice the stiffening power of beeswax.

CHAPTER 6

Body Beautiful

Bath and body products have always been an obsession for me. I derive immense joy from blending vegetable oils loaded with nutrients, intensely aromatic and therapeutic botanical oils, an array of mineral-rich salts, and many other natural ingredients to enrich, nourish, soothe, or invigorate the skin. Their effects are delivered in many ways: inhaled, applied topically, or—best of all—absorbed in the steamy and relaxing comfort of a hot bath.

Gentle Body Cleanser

The combination of sweet agave nectar, rich coconut milk, and effective yet tender castile soap creates a skin-kind body cleanser suitable for dry and sensitive skin. The aromatic combination of ginger, orange, and lemon essential oils produces a warming, invigorating, and encouraging scent, which will fill the air in your bathroom for hours after your shower or bath. A heavenly way to kick-start your day!

5 tablespoons castile soap

2 tablespoons glycerine

2 teaspoons agave nectar

3 teaspoons coconut milk

1 teaspoon CBD oil

5 drops ginger essential oil

5 drops orange essential oil

5 drops lemon essential oil

Equipment

Glass or stainless-steel beaker

Small hand whisk

Glass or wooden stirrer

4-fl oz (120-ml) glass bottle with flip cap or pump dispenser

Preparation time: 10 minutes

Makes 4 fl oz (120 ml)

1 Combine the castile soap, glycerine, and agave nectar in the beaker and mix with the hand whisk.

2 Add the coconut milk and whisk until all ingredients are incorporated.

3 Slowly stir in the CBD oil along with the ginger, orange, and lemon essential oils.

4 Transfer the mixture to the glass bottle and seal.

STORAGE

This cleanser will keep for up to two weeks if stored in a cool place, preferably the refrigerator.

TO USE

Massage into your body during a shower or bath for a soothing cleansing experience. Oil-based products can make the shower or bathtub floor slippery, so be sure to use a nonslip mat.

APOTHECARY NOTE

Castile soap is a brilliantly useful and convenient base ingredient for homemade beauty recipes. True castile soap is made from olive oil (as opposed to poorer-quality castile soap made from synthetic compounds), which has a high concentration of essential fatty acids, making it a very gentle cleansing ingredient. Unlike traditional soap, which contains animal fats, castile soap is vegan and cruelty-free.

Cucumber and Rosemary Body Scrub

Our body sheds dead skin cells every three weeks or so to make room for new cells. As we grow older, or become exposed to environmental stresses, this natural regenerative process slows down—resulting in flaky patches, which dull the skin and clog the pores. The greatest gift we can give our body is regular whole-body exfoliation. This body scrub harnesses the cooling and refreshing properties of cucumber and rosemary, along with the powerful antioxidant capacity of CBD oil. Your skin will thank you as it becomes visibly brighter and clearer!

½ cucumber (around 7 oz/200 g), seeded, peeled, and chopped

1 small bunch rosemary, stems removed

1 small bunch thyme, stems removed

2 tablespoons Himalayan salt

½ teaspoon CBD oil

2 drops rosemary essential oil

Equipment

Standard kitchen blender

Glass or stainless-steel bowl

Glass or wooden stirrer

Preparation time: 10 minutes

Makes 8 oz (225 g) for single immediate use

1 Blend the cucumber, rosemary, and thyme for 1–2 minutes, until the ingredients are well combined and foamy in texture.

2 Transfer the mixture to the bowl and stir in the Himalayan salt, CBD oil, and rosemary essential oil.

TO USE

This recipe should be used as soon as it is made—the fresher the better! Once in the shower or bathtub, firmly massage the scrub into your body using circular upward movements for a refreshed and polished feeling. Use it weekly to support your skin's own cell regeneration process. Be sure to use a nonslip bath or shower mat.

APOTHECARY NOTE

Himalayan salt scrubs are a fabulous exfoliator, but they can be a little too powerful for super-sensitive skin. If this is the case with you, replace the salt with brown sugar for gentler but still very effective exfoliation.

Coffee Body Scrub

I have been a fan of coffee scrubs for years. Now, CBD oil and coffee have become a power pairing in the world of food, so it feels natural to combine these two ingredients to promote our skin health. Not only is the scent of coffee delicious, but caffeine is an amazing antioxidant with impressive skin cell regenerative properties. In this scrub, the coffee works away at its polishing task, while CBD sweeps in to soothe and calm your skin. Once you have tried this, it is sure to become a staple of your beauty routine.

3 tablespoons chickpea flour

1 tablespoon organic brown sugar

1 tablespoon ground coffee

3 teaspoons ground hempseeds

2 teaspoons shea butter

1 tablespoon hempseed oil

1 teaspoon CBD oil

1 teaspoon vanilla extract

4 drops grapefruit essential oil

Equipment

2 glass or stainless-steel bowls

Glass or wooden stirrer

Small hand whisk

2-oz (55-g) glass jar with lid

Preparation time: 5 minutes

Makes 2 oz (55 g)

1 In the first bowl, combine all the dry ingredients—the chickpea flour, brown sugar, ground coffee, and ground hempseeds.

2 In the second bowl, whip the shea butter with the hand whisk until smooth. Slowly whisk in the hempseed oil, CBD oil, vanilla extract, and grapefruit essential oil.

3 Add the dry ingredients to the butter mixture and continue to stir until fully incorporated.

4 Transfer the mixture to the jar and seal.

STORAGE

The scrub will keep for two weeks.

TO USE

Apply generously to wet or dry skin and massage in a circular motion to encourage exfoliation. Rinse thoroughly.

APOTHECARY NOTES

Weekly exfoliation is a critical element of your skincare routine. If you allow old skin cells to accumulate on the surface of your skin, excess oil will build, your pores will become clogged, your moisturizer will not penetrate efficiently, and your skin will be dull and lackluster.

Feel free to use either fresh coffee or leftover coffee grounds in this recipe—either will work beautifully.

Vanilla-enriched Body Butter

Body butters are waterless moisturizers which offer a particularly hydrating experience for your skin. Cocoa butter is one of my favorite ingredients in this CBD-infused body butter as it is a powerful antioxidant, packed with vitamin E and fatty acids. The result is a truly impressive hydrating and nourishing impact. The presence of vanilla extract and essential oils in this butter is the icing on the cake, leaving the most exquisite scent on your skin!

2 oz (55 g) shea butter

2 oz (55 g) cocoa butter

3 tablespoons baobab oil

3 tablespoons argan oil

2 teaspoons CBD oil

½ teaspoon vanilla extract

15 drops grapefruit essential oil

10 drops frankincense essential oil

5 drops ylang ylang essential oil

5 drops Palo Santo essential oil

Equipment

Double boiler

Glass stirrer

Hand mixer with whisk attachment, or a hand whisk

Stainless-steel bowl

Spatula

8-oz (225-g) glass jar with lid

Preparation time: 90 minutes

Makes 8 oz (225 g)

1 Melt the shea butter and cocoa butter in the double boiler.

2 Remove the bowl from the heat and stir in the baobab, argan, and CBD oils. Let the mixture cool for 10 minutes.

3 Whisk the mixture for 2–3 minutes on a low speed until it starts to become creamy in texture, then refrigerate for 1 hour.

4 Remove from the refrigerator and whisk again for 3 minutes, stopping occasionally to scrape the mixture from the sides of the bowl into the center with a spatula.

5 Gently stir in the vanilla extract and grapefruit, frankincense, ylang ylang, and Palo Santo essential oils, then whisk the mixture for another three minutes.

6 Transfer the mixture to the glass jar and seal.

STORAGE

As this body butter contains no water and is packed with antioxidant ingredients, it will keep for six months.

TO USE

Massage a generous amount of the butter into your skin after bathing or showering. This body butter can be used daily and is especially appreciated by dry skin.

Frankincense Nourishing Body Oil

Where to begin in describing the supremely aromatic king of oils? Frankincense has been used since the most ancient times for medicinal and wellness purposes. It was considered more valuable than gold and was burned as an offering to the gods. Its fragrance promotes feelings of calm, relaxation, and wellbeing, and in the beauty world the oil is prized for its ability to regenerate and rejuvenate the skin. This recipe partners frankincense with vitamin E-rich peach kernel oil and hempseed oil, which is brimming with omega-3, -6, and -9. For whole-body support, it is a winning combination.

2 tablespoons olive oil

1 tablespoon hempseed oil

1 teaspoon peach kernel oil

1 teaspoon CBD oil

10 drops cedarwood essential oil

6 drops frankincense essential oil

3 drops orange essential oil

Equipment

Glass or stainless-steel beaker

Glass or wooden stirrer

2-fl oz (60-ml) glass bottle with dropper cap or flip lid

Preparation time: 5 minutes

Makes 2 fl oz (60 ml)

1 Combine the olive, hempseed, peach kernel, and CBD oils in the beaker.

2 Gently stir in the cedarwood, frankincense, and orange essential oils.

3 Transfer the mixture to the glass bottle and seal.

STORAGE

As there is no water in this body oil, it will keep for up to three months if stored away from direct sunlight and heat.

TO USE

Generously massage into your skin after your daily shower or bath. I especially adore this body oil at night. The frankincense is the perfect resinous aroma to float through your senses just before bed.

APOTHECARY NOTE

Hempseed and peach kernel carrier oils are deeply penetrative. As they reach down through the layers of the skin, they carry with them the essential oils, enabling their absorption at the deepest level.

Rose Petal Vegan Bath Milk

A bath is one of the greatest gifts we can give ourselves. Telling the world that you deserve to soak your beautiful body for 20 sweet minutes is a gorgeous act of self-love! Adding some therapeutic and aromatic ingredients to that bath is obligatory. This recipe is a vegan alternative to dairy. With its cocoa butter, CBD oil, and heavenly geranium and ylang ylang essential oils, it is no less luxurious and emollient!

5½ oz (150 g) chickpea flour

Vanilla pod

Handful of rose petals, fresh or dried

2 oz (55 g) cocoa butter, chilled

3 teaspoons CBD oil

10 drops geranium essential oil

10 drops ylang ylang essential oil

Equipment

Glass or stainless-steel bowl

Standard kitchen grater

Glass or wooden stirrer

8-oz (225-g) glass jar with lid

Preparation time: 5 minutes

Makes 8 oz (225 g)

1 Combine the chickpea flour, vanilla pod, and rose petals in the bowl.

2 Grate the chilled cocoa butter into the bowl.

3 Stir in the CBD oil and the geranium and ylang ylang essential oils.

4 Transfer to the glass jar and seal.

STORAGE

This recipe will keep for up to a year unopened, but use it up within six weeks once the seal is broken. The longer you leave this bath milk sealed in the jar, the more aromatically infused the mixture will become. The fragrant scent from the jar once you open it will be stunningly beautiful and the bathing experience is one of the most luxurious you will ever have.

TO USE

While you are running your bath, add three scoops of the bath milk. Swirl it around with your hand to ensure even distribution. If you are feeling very Cleopatra-like, add an additional scoop for a milkier bath! Be sure to use a non-slip bathmat.

APOTHECARY NOTE

Never heard of chickpea flour and certainly never in a skincare recipe? Don't worry—it is not a crazy concept! Chickpea flour will soothe and smooth your skin, leaving you emerging from your bath with baby-soft skin.

Aromatherapy Bath Teabags

Herbal bath soaks are wonderfully relaxing. However, they can be very messy! The answer is to make your own bath "teabags" using muslin squares. That way, you have all the benefits of the herbs with none of the debris on your skin and bathtub.

2 tablespoons Dead Sea salts

1 tablespoon dried lavender flowers

1 tablespoon dried rose petals

1 tablespoon dried nettle herb

1 tablespoon dried seaweed

1 teaspoon CBD oil

10 drops lavender essential oil

Equipment

Glass or stainless-steel bowl

Glass or wooden stirrer

4 x 8-in. (20-cm) squares of muslin

4 elastic bands

4 lengths of string or ribbon

Preparation time: 5 minutes

Makes four teabags

1 Combine the Dead Sea salts, lavender flowers, rose petals, nettle herb, and seaweed in the bowl.

2 Stir in the CBD oil and lavender essential oil.

3 Spoon the mixture evenly into the four muslin squares.

4 Gather the corners of each muslin square and secure with an elastic band.

5 Tie the teabags with string—or ribbon, to make them oh so pretty!

STORAGE

The teabags will keep for up to 12 months.

TO USE

Place one teabag into your bathtub and relax!

Palo Santo Sleep Bath Oil

Palo Santo ("Holy Wood") originated in South America, where it was used by the Incas in their rituals to purify the spirit of negative energy and to achieve better spiritual communication with their gods. Today, Palo Santo essential oil is often used in yoga and other meditative practices as it promotes a calm and peaceful atmosphere. A relative newcomer to the skincare scene, Palo Santo is valued for its anti-inflammatory properties.

2 tablespoons jojoba oil

1½ tablespoons grapeseed oil

1 teaspoon CBD oil

7 drops frankincense essential oil

5 drops lavender essential oil

5 drops Roman chamomile essential oil

2 drops Palo Santo essential oil

Equipment

Glass or stainless-steel beaker

Glass or wooden stirrer

2-fl oz (60-ml) glass bottle with dropper cap or flip lid

Preparation time: 5 minutes

Makes 2 fl oz (60 ml)

1 Combine the jojoba, grapeseed, and CBD oils in the beaker.

2 Gently stir in the frankincense, lavender, Roman chamomile, and Palo Santo essential oils.

3 Transfer the mixture to the glass bottle and seal.

STORAGE

As there is no water in this Palo Santo sleep bath oil, it will keep for up to three months if stored away from direct sunlight and heat.

TO USE

Add a generous pour to your bathtub as you run the water and luxuriate in the scent as it fills the bathroom. Be sure to use a non-slip bathmat.

APOTHECARY NOTE

Palo Santo has a strong aroma and is effective in a low dose, hence the small quantity in this recipe. The frankincense fragrance complements that of Palo Santo beautifully. Marrying these two essential oils in a diffuser can greatly enhance any yoga or meditation practice.

Lavender Sleep Massage Bar

Lavender has long been the go-to essential oil for promoting restful sleep. Here it is joined by CBD oil to form a highly effective sleep support. Pain, stress, and anxiety are three of the major causes of sleep disturbance, and CBD oil works effectively on all three fronts. Blending these formidable sleep aids in one recipe creates a truly blissful tool for encouraging a peaceful night between the sheets.

1 oz (30 g) cocoa butter

½ oz (15 g) shea butter

1 teaspoon CBD oil

10 drops lavender essential oil

10 drops cedarwood essential oil

10 drops frankincense essential oil

1 teaspoon dried lavender flowers

Equipment

Double boiler

Stainless-steel bowl

Glass stirrer

2 x 1-oz (30-g) molds or 6 smaller paper muffin/cupcake cases

Preparation time: 20 minutes

Makes two 1-oz (30-g) molds, but this recipe can yield up to six small bars if you choose to use small bun cases instead of larger molds

1 Melt the cocoa butter and shea butter in the double boiler.

2 Remove from the heat and gently stir in the CBD oil and lavender, cedarwood, and frankincense essential oils, followed by the lavender flowers.

3 Pour the mixture into the molds and place in the refrigerator to set for 1–2 hours.

4 Once the mixture has set, peel the massage bars from the molds.

STORAGE
The massage bars will keep for up to a year if stored somewhere cool.

TO USE
Massage into the body as needed before going to sleep.

APOTHECARY NOTE
The base of shea and cocoa butters in these massage bars allows a smooth massage application. They nourish and condition as you work the aromatic sleep-inducing plant nutrients into your skin.

CHAPTER 7

Hair, Hands, and Feet

They may be our extremities, but our hair, hands, and feet are arguably our hardest-working assets. They carry us, protect us, and face the world for us. They deserve, but do not always get, the same levels of tender loving care that we lavish on our face and body. Infusing our hair and scalp with plant-based vitamins and nutrients stimulates growth and improves volume, shine, and texture. Our hands and feet, which work so tirelessly for us, require nourishment and CBD oil delivers gentle support to both the skin and muscles.

Tea Tree Hand Sanitizing Gel

Hand sanitizers are now part of our lives. However, there is no denying that alcohol-based sanitizers are hard on our skin! Here is a gentle alternative, drawing on the hydrating activity of aloe vera and the antimicrobial and antifungal properties of tea tree and witch hazel.

1 tablespoon aloe vera gel

2 teaspoons witch hazel

½ teaspoon CBD oil

5 drops tea tree essential oil

5 drops lemon essential oil

5 drops orange essential oil

5 drops cedarwood essential oil

Equipment

Glass or stainless-steel beaker

Glass or wooden stirrer

1-fl oz (30-ml) glass bottle with pump application

Preparation time: 5 minutes

Makes 1 fl oz (30 ml)

1 Combine the aloe vera gel, witch hazel, and CBD oil in the beaker.

2 Gently stir in the tea tree, lemon, orange, and cedarwood essential oils.

3 Transfer the mixture to the glass bottle and seal.

STORAGE

The antimicrobial properties of tea tree essential oil will support the stability of this recipe. However, it is recommended that you use it within one month.

APOTHECARY NOTE

This gel, with its lemon, orange, and cedarwood oils, has a pleasant citrus aroma with a hint of wood—an agreeable contrast to most alcohol-based sanitizing gels.

Chickpea Hand Mask

We regularly exfoliate and moisturize our face and body, but how often do we extend that level of care to one of our most hardworking and exposed body parts? This luscious mask is the kindest of treatments for our often-neglected hands, which will emerge polished, hydrated, and so smooth. An added benefit is that you are forced to sit still for ten minutes while the mask works its alchemy.

1 teaspoon argan oil

1 teaspoon agave nectar

1 teaspoon lavender hydrosol

½ teaspoon CBD oil

1 teaspoon white clay

½ teaspoon chickpea flour

½ teaspoon oat flour

½ teaspoon dried nettle herb

4 drops ylang ylang essential oil

2 drops geranium essential oil

2 drops carrot seed essential oil

Equipment

Glass or stainless-steel bowl

Glass or wooden stirrer

Preparation time: 5 minutes

Makes enough for single immediate use

1 Combine the argan oil, agave nectar, lavender hydrosol, and CBD oil in the bowl.

2 Slowly add the white clay, chickpea flour, oat flour, and dried nettle herb.

3 Stir in the ylang ylang, geranium, and carrot seed essential oils.

4 Stir the mixture until the ingredients are fully incorporated.

TO USE

Dampen your hands with water or a botanical hydrosol. Smother your hands with the mask, thoroughly covering your fingers. Sit still for 10 minutes and breathe gently. The mask will harden because of the white clay. When it starts to crack, rinse it off with warm water and use a cloth to remove any residue.

APOTHECARY NOTE

If you have a botanical hydrosol, use it instead of water to dampen your hands before application. Such hydrosols smell divine and add to the therapeutic properties of this hand mask.

Peppermint Foot Scrub

Our sweet feet deserve much care as they literally carry us all day. The menthol in this oil will create a cooling and satisfyingly tingly sensation which will leave your hard-working soles refreshed and invigorated.

2 tablespoons organic brown sugar

1 tablespoon mustard seed powder

1 tablespoon hempseed oil

1 teaspoon castor oil

1 teaspoon CBD oil

6 drops peppermint essential oil

4 drops ginger essential oil

Equipment

Glass or stainless-steel bowl

Glass or wooden stirrer

2-oz (55-g) glass jar with lid

Preparation time: 5 minutes

Makes 2 oz (55 g)

1 Combine the organic brown sugar and mustard seed powder in the bowl.

2 Stir in the hempseed, castor, and CBD oils.

3 Add the peppermint and ginger essential oils and stir until all the ingredients are fully integrated.

4 Transfer the mixture to the glass jar and seal.

STORAGE

As there is no water in this peppermint foot scrub, it will keep for up to three months if stored away from direct sunlight and heat.

TO USE

Take a generous amount of the peppermint foot scrub between your palms. Massage into your feet until they feel smooth and polished. Rinse with warm water and use a cloth to remove any residue.

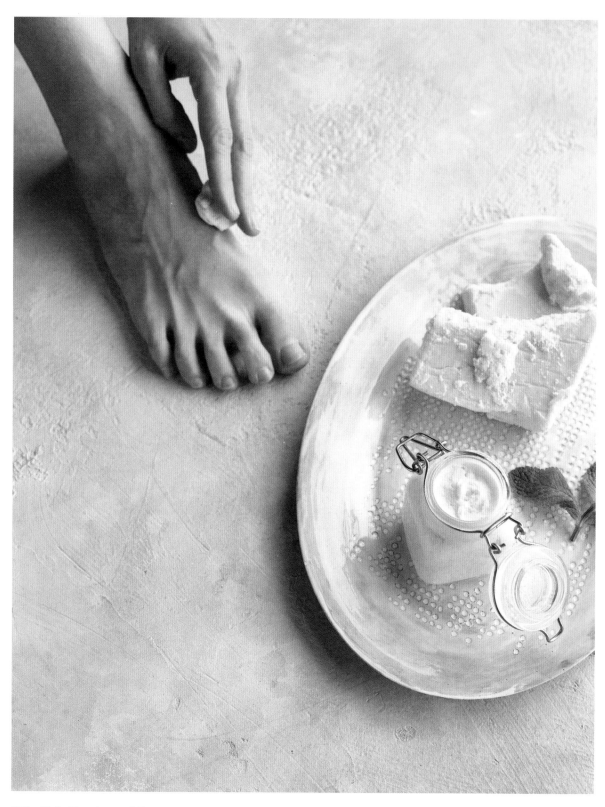

Luxurious Foot Cream

Foot massage can really get on your nerves—the 7,000 nerve endings in your feet, that is! This super-hydrating and stimulating recipe combines rich and creamy shea butter with an infusion of CBD and botanical oils. Work the cream deeply into each foot until it is completely absorbed. Not only will your feet feel wonderful, but the rest of your body will thank you, too.

3 tablespoons shea butter

1 teaspoon argan oil

1 teaspoon hempseed oil

1 teaspoon CBD oil

10 drops peppermint essential oil

4 drops cedarwood essential oil

Equipment

Hand mixer with whisk attachment, or a hand whisk

Glass or stainless-steel bowl

Spatula

2-oz (55-g) glass jar with airtight lid

Preparation time: 15 minutes

Makes 2 oz (55 g)

1 Whip the shea butter with the hand mixer on high for three minutes, until it begins to take on a creamy texture.

2 Gradually add the argan, hempseed, and CBD oils, followed by the peppermint and cedarwood essential oils, while continuing to whip the mixture until the oil is fully incorporated into the shea butter and the mixture is rich and creamy.

3 Transfer to the glass jar using the spatula, and seal.

STORAGE

This product will last for two months if you store it in an airtight container and keep it away from direct sunlight and heat.

TO USE

Take a generous amount of the luxurious foot cream and massage into your feet until they feel beautifully moisturized.

APOTHECARY NOTE

Show your feet some extra tender loving care by using this luxurious cream as an overnight foot mask. Simply apply an extra-generous amount at night and pull on a pair of socks before you go to bed. Your feet will emerge from those socks in the morning softer than ever before.

Black Cumin Scalp Serum

Black cumin is one of the oldest herbs known to mankind. In the Arab world, it is called the "Seed of Blessing" (the Prophet Muhammed declared that it cured all diseases except death). It is a powerhouse of healthy fatty acids with antioxidant, antibacterial, and antihistamine properties. Recent studies indicate that, applied topically to the scalp, black cumin oil strengthens hair follicles and promotes improved hair growth, volume, and texture.

2 teaspoons jojoba oil

1 teaspoon argan oil

1 teaspoon rice bran oil

1 teaspoon black cumin oil

1 teaspoon CBD oil

2 drops ylang ylang essential oil

2 drops cedarwood essential oil

2 drops petitgrain essential oil

Equipment

Glass or stainless-steel beaker

Glass or wooden stirrer

1-fl oz (30-ml) glass bottle with dropper cap or flip lid

Preparation time: 5 minutes

Makes 1 fl oz (30 ml)

1 Combine the jojoba, argan, rice bran, black cumin, and CBD oils in the beaker.

2 Gently stir in the ylang ylang, cedarwood, and petitgrain essential oils.

3 Transfer the mixture to the glass bottle and seal.

STORAGE

As there is no water in this serum, it will keep for up to two months if stored away from direct sunlight and heat.

TO USE

Take a generous amount of the serum and massage thoroughly into your scalp for 5 minutes. Wrap your head in a heated towel and relax for 15 minutes to allow the serum time to penetrate and nourish your scalp. Finish by washing your hair with a gentle shampoo.

APOTHECARY NOTE

Don't use a blow dryer after this exquisitely therapeutic treatment. Allow your hair to dry naturally instead.

Hempseed Hair Serum

Hempseed and sesame oils are similar in that they are both rich in omega-3, -6, and -9, which collectively provide a protective and balancing barrier for our hair and skin. They offer a restorative and hydrating experience. Wonderful penetration abilities intensify the therapeutic effect of these marvelous oils.

2 teaspoons sesame oil

1 teaspoon argan oil

1 teaspoon hempseed oil

1 teaspoon peach kernel oil

1 teaspoon CBD oil

2 drops peppermint essential oil

3 drops rosemary essential oil

2 drops lavender essential oil

Equipment

Glass or stainless-steel beaker

Glass or wooden stirrer

1-oz (30-ml) glass bottle with dropper cap or flip lid

Preparation time: 5 minutes

Makes 1 oz (30 ml)

1 Combine the sesame, argan, hempseed, peach kernel, and CBD oils in the beaker.

2 Gently stir in the peppermint, rosemary, and lavender essential oils.

3 Transfer the mixture to the glass bottle and seal.

STORAGE

As there is no water in this hair serum, it will keep for up to two months if stored away from direct sunlight and heat.

TO USE

Take a generous amount of the serum and massage through your hair for 5 minutes. Wrap your head in a heated towel and rest for 15 minutes to allow the serum time to penetrate and nourish your hair and scalp. Finish by washing your hair with a gentle shampoo.

CHAPTER 8

Supported by CBD

CBD oil's formidable anti-inflammatory and antioxidant capabilities position it as the core plant ingredient in our CBD support recipes. These everyday recipes help with pain relief (chronic and acute), promoting peaceful and restorative sleep, and boosting our immune system.

Arnica Muscle Massage Oil

CBD, arnica, comfrey, and trauma oils are a remarkable combination of carrier oils with significant anti-inflammatory and pain-relieving properties. They are especially beneficial in treating muscular and skeletal tissue problems—including sprains, pulled muscles, muscle spasms, and neck tension, as well as bruises and swelling. Complemented with essential oils, this massage oil is rich in properties which stimulate your body's own powers of regeneration and healing.

2 tablespoons sesame oil

2 teaspoons arnica oil

1 teaspoon argan oil

1 teaspoon comfrey oil

1 teaspoon trauma oil

1 teaspoon CBD oil

5 drops black pepper essential oil

5 drops ginger essential oil

5 drops peppermint essential oil

5 drops lavender essential oil

3 drops marjoram essential oil

Equipment

Glass or stainless-steel beaker

Glass or wooden stirrer

2-fl oz (60-ml) glass bottle with dropper cap or flip lid

Preparation time: 5 minutes

Makes 2 fl oz (60 ml)

1 Combine the sesame, arnica, argan, comfrey, trauma, and CBD oils in the beaker.

2 Gently stir in the black pepper, ginger, peppermint, lavender, and marjoram essential oils.

3 Transfer the mixture to the glass bottle and seal.

STORAGE

As there is no water in this massage oil, it will keep for up to six months if stored away from direct sunlight and heat.

TO USE

Take a generous amount of the oil in your hands and rub into the area of pain as required.

APOTHECARY NOTE

If you are short on time, or simply prefer a soak as opposed to a massage, this blend doubles up as a lovely and effective bath oil. Add a generous pour into the bathtub as you run the water, and soak those aching muscles. As with all oil-based products, use a non-slip bathmat.

Pain Relief Gel

CBD oil is known for its analgesic properties, helping ease pain, inflammation, and discomfort. Aloe vera not only provides effective anti-inflammatory action, but its rapid absorption by the skin accelerates the pain-relieving properties of this useful blend.

1 tablespoon aloe vera gel

1 teaspoon lavender hydrosol

1 teaspoon CBD oil

10 drops black pepper essential oil

8 drops peppermint essential oil

8 drops lavender essential oil

Equipment

Glass or stainless-steel beaker

Glass or wooden stirrer

1-fl oz (30-ml) glass bottle with dropper cap or flip lid

Preparation time: 5 minutes

Makes 1 fl oz (30 ml)

1 Combine the aloe vera gel, lavender hydrosol, and CBD oil in the beaker.

2 Gently stir in the black pepper, peppermint, and lavender oils.

3 Transfer to the glass bottle and seal.

STORAGE

This gel will remain fresh for up to one month if kept refrigerated.

TO USE

Shake well before use, then massage into the affected area.

Candelilla Pain-relieving Salve

**While oil-based products are very good at delivering speedy relief in cases
of short-lived acute pain, a wax-based salve (in this vegan recipe we use
candelilla wax) promotes slow-release analgesia for people with chronic pain,
such as arthritis.**

2 tablespoons candelilla wax

1 tablespoon olive oil

1 tablespoon hempseed oil

1 teaspoon trauma oil

1 teaspoon CBD oil

5 drops black pepper essential oil

5 drops ginger essential oil

5 drops rosemary essential oil

5 drops Roman chamomile essential oil

5 drops marjoram essential oil

Equipment

Double boiler

Glass stirrer

2 oz (55 g) glass jar with lid

Preparation time: 5 minutes

Makes 2 oz (55 g)

1 Melt the candelilla wax in the double boiler.

2 Keeping the bowl over the heat, stir in the olive, hempseed, trauma, and CBD oils.

3 Remove from the heat and gently stir in the black pepper, ginger, rosemary, Roman chamomile, and marjoram essential oils.

4 Transfer the mixture to the glass jar and seal.

STORAGE

As there is no water in this salve, it will keep for up to six months if stored away from direct sunlight and heat.

TO USE

Apply the pain-relieving salve to areas of stiffness and tension as frequently as required.

Muscle-soothing Bath Salts

In this recipe, we boost CBD oil's analgesic and anti-inflammatory properties with magnesium flakes and Epsom salts. Magnesium is a vitally important mineral, yet it is estimated that 68% of Americans have a magnesium deficiency. Low magnesium intake is linked to chronic inflammation, which is a major source of pain and one of the drivers of aging. Interestingly, one of the most effective ways of absorbing magnesium is through your skin rather than via your digestive system—so a therapeutic soak is thoroughly recommended for those tired and sore muscles!

6 oz (170 g) Epsom salts

6 oz (170 g) magnesium flakes

3 oz (85 g) Dead Sea salts

1 tablespoon matcha powder

1 tablespoon CBD oil

20 drops rosemary essential oil

20 drops frankincense essential oil

20 drops cedarwood essential oil

Equipment

Glass or stainless-steel bowl

Glass or wooden stirrer

1-lb (450-g) glass jar with lid

Preparation time: 5 minutes

Makes 1 lb (450 g)

1 Combine the Epsom salts, magnesium flakes, and Dead Sea salts in the bowl.

2 Add the matcha powder and stir until the entire mixture turns a shade of light green.

3 Stir in the CBD oil and rosemary, frankincense, and cedarwood essential oils.

4 Transfer the mixture to the glass jar and seal.

STORAGE

This large jar will last up to four months. Keep it sealed when not in use to avoid moisture getting into the salts.

TO USE

Scoop two heaped tablespoons of the muscle-soothing salts into the bathtub as you run the water. Add a little extra if your body is tired or if you have been overworking those muscles. Be sure to use a non-slip bathmat.

APOTHECARY NOTE

Make sure to clean your bathtub well immediately after use as the matcha can be difficult to remove once it dries.

Overnight Immune-boost Massage Oil

Massage induces a relaxation response in our body, and at the same time it calms and quietens our mind. When the massage is delivered with this marvelous overnight blend, bursting with nutritious immunostimulant plant oils, you have the perfect pairing. While you enjoy deeper, more peaceful sleep, your body will absorb a cocktail of nature's delicious and restorative aromatic oils. What's not to enjoy?

2 tablespoons jojoba oil

1 tablespoon hempseed oil

2 teaspoons CBD oil

10 drops lemon essential oil

10 drops tea tree essential oil

10 drops frankincense essential oil

10 drops petitgrain essential oil

10 drops Roman chamomile essential oil

Equipment

Glass or stainless-steel beaker

Glass or wooden stirrer

2-fl oz (60-ml) glass bottle with lid

Preparation time: 5 minutes

Makes 2 fl oz (60 ml)

1 Combine the jojoba, hempseed, and CBD oils in the beaker.

2 Gently stir in the lemon, tea tree, frankincense, petitgrain, and Roman chamomile essential oils.

3 Transfer the mixture to the glass bottle and seal.

STORAGE

As there is no water in this massage oil, it will keep for up to three months if stored away from direct sunlight and heat.

TO USE

Massage into the body as often as required and especially before sleep as the perfect pre-bed therapy.

Sweet Dreams Pillow Spray

Infuse your pillow with heavenly lavender and relaxing CBD oil to ensure you have sweet, sweet dreams all night long!

4 teaspoons lavender hydrosol

1 teaspoon glycerine

1 teaspoon CBD oil

2 drops lavender essential oil

Equipment

Glass or stainless-steel beaker

Glass or wooden stirrer

1-fl oz (30-ml) glass bottle with spray application top

Preparation time: 5 minutes

Makes 1 fl oz (30 ml)

1 Combine the lavender hydrosol, glycerine, and CBD oil in the beaker.

2 Gently stir in the lavender essential oil.

3 Transfer to the glass bottle and seal.

STORAGE

Store away from direct sunlight and heat. Your spray will stay fresh and aromatic for up to three weeks.

TO USE

Spray directly onto your pillow before sleep.

Suppliers

The beauty of this book is that no recipe requires special equipment. Nor do any of the recipes require inaccessible ingredients that you can't find in your local health or grocery store, or online.

On that note, let's talk about suppliers. Having reputable suppliers is half the battle when making CBD beauty products in your kitchen. In this section, I share my own treasure trove of suppliers: some I have loved for years, others are more recent favorites. I have included global suppliers to help you find the best options no matter where you are.

Before you go on a wild shopping spree for all these stunning plant ingredients and fill your kitchen with endless cute glass beakers, let me give you a shopping tip. I default to natural and, where possible, organic ingredients when blending products for my skin and body. Don't be surprised when you end up simply extending your grocery or health store shopping list as opposed to putting in large orders with cosmetic or aromatherapy ingredient suppliers.

CBD SOURCES

www.cbii-cbd.com (UK/EU)

www.rosebudcbd.com (US)

www.shop-poplar.com (US/UK/EU/CA/ROW)

www.bloomfarmscbd.com (US)

www.charlottesweb.com (US)

www.blacktiecbd.net (US)

www.love-hemp.com (UK/EU)

www.cbd-guru.co.uk (UK/EU/ROW)

www.junepure.com (EU/UK)

CBD AND ESSENTIAL OILS LEARNING RESOURCES

www.labaroma.com/en/training

www.aromatics.com/pages/learning-guides

www.projectcbd.org

CONTAINERS

www.ampulla.co.uk (UK)

www.lesamesfleurs.com (CA)

www.stocksmetic.com (US/UK/CA)

www.amazon.com (UK/US/EU/CA/ROW)

LABELS

www.onlinelabels.com (UK/US)

www.a4labels.com (EU/UK/US/CA)

ESSENTIAL OILS

www.aromatics.com (US/UK/EU/CA/ROW)

www.oshadhi.co.uk (UK/EU)

www.newdirectionsaromatics.ca (US/CA)

www.escents.ca (US/CA)

www.edensgarden.com (US/UK/EU/CA/ROW)

www.baseformula.com (UK)

www.nealsyardremedies.com (UK/EU)

CARRIER OILS

www.aromatics.com (US/UK/EU/CA/ROW)

www.baldwins.co.uk (UK/EU)

www.nobleroots.com (US)

www.baseformula.com (UK)

BUTTERS AND WAXES

www.aromatics.com (US/UK/EU/CA/ROW)

www.hollandandbarrett.com (UK/EU)

www.thesoapkitchen.co.uk (UK/EU)

www.mountainroseherbs.com (US/CA)

www.organic-creations.com (US)

www.baseformula.com (UK)

HYDROSOLS

www.aromatics.com (US/UK/EU/CA/ROW)

www.edenbotanicals.com (US/CA)

www.mountainroseherbs.com (US/CA)

CLAYS AND SALTS

www.realsalt.com (US)

www.seasalt.com (US/CA)

www.sunrisebotanics.com (US/CA)

www.bayhousearomatics.com (US/UK/EU/ROW)

www.mountainroseherbs.com (US/CA)

www.baldwins.co.uk (UK/EU)

EQUIPMENT

www.walmart.com (US)

www.crateandbarrel.com (US/UK)

www.makingcosmetics.com (US)

www.zedmed.co.uk (UK/EU)

www.ikea.com (UK/EU)

References

Page 14: 1. Quoted in "Cannabinoid Delivery Systems for Pain and Inflammation Treatment" by Natascia Bruni, Carlo Della Pepa, Simonetta Oliaro-Bosso, Enrica Pessione, Daniela Gastaldi, and Franco Dosio, *Molecules*, October 2018.

Page 15: 2. Quoted in *Road to Ananda* by Carl Germano (Healthy Living Publishing LLC, January 2019)

Page 15: 3. Quoted in *Healing with CBD* by Eileen Konieczny and Lauren Wilson (Ulysses Press, October 2018)

Picture Credits

All photography by Joanne Gregory © CICO Books unless otherwise stated below.

Page 5: Stuart West © CICO Books

Page 12: Clare Winfield © Ryland Peters and Small

Page 17: © mentWorks/Shutterstock.com

Page 20: Clare Winfield © Ryland Peters and Small

Page 22: Kate Whitaker © Ryland Peters and Small

Page 24: © solar22/Shutterstock.com

Page 26: © Ryland Peters and Small

Page 27: Steve Painter © Ryland Peters and Small

Page 30: Clare Winfield © Ryland Peters and Small

Page 32: Peter Cassidy © Ryland Peters and Small

Page 33: Dan Duchars © Ryland Peters and Small

Page 35: Stuart West © CICO Books

Page 39: Kate Whitaker © Ryland Peters and Small

Page 40: Richard Jung © Ryland Peters and Small

Page 41: © Ryland Peters and Small

Page 43: Kate Whitaker © Ryland Peters and Small

Page 45: Kate Whitaker © Ryland Peters and Small

Page 46: Erin Kunkel © Ryland Peters and Small

Page 47: Peter Cassidy © Ryland Peters and Small

Page 48: Stuart West © CICO Books

Page 49: Mowie Kay © Ryland Peters and Small

Page 50: Stuart West © CICO Books

Page 51: Kate Whitaker © Ryland Peters and Small

Page 52: Toby Scott © Ryland Peters and Small

Page 55: Clare Winfield © Ryland Peters and Small

Page 58: Stuart West © CICO Books

Index

Acknowledgments

A wise woman once told me that anything worth doing will take a village to accomplish. Thankfully my village is packed with talented and dedicated people who are as responsible for this book as I am.

To Helen, my endlessly patient editor. Thank you for pushing me, challenging me, and making me a better writer, teacher, and creator. Your patience is the greatest gift you give to me as we construct masterworks together.

To Emma and Jo, who so beautifully brought these recipes to life.

To Carmel and the team at CICO, thank you for your confidence in me and for giving me a blank page with which to share my CBD and botanical knowledge.

As I wrote this book, my team at LabAroma was deep in the eleventh hour of our greatest project yet. It is thanks to Sarah for taking the baton from me and leading so perfectly that this book is in your hands and LabAroma continues to grow and evolve to serve a wider plant-loving community. To Sarah, Enya, Jonathan, Mary Joy, Sean, Caterina, Monique, Prachanti, Amalthea, Brian, and Niall—thank you for your dedication and commitment to our mission. The LabAroma community are the most fortunate group of learners to have this devoted team supporting them. Thanks also to Sheila, Christina, and Lisa, the creative queens who turn my words and thoughts into gorgeous, accessible projects.

To my LabAroma community, thank you for your support and loyalty. Our weekly calls have become indispensable to me as you turn up to direct me while I teach and curate software products to make your plant-blending experiences smoother, innovative, informed, and, most importantly, more fun. Thank you for always communicating your greatest dreams and making sure I deliver them to you as you wished. I am eternally humbled by the trust you place in me.

There is a group of women who have taught me as much about the intricacies of life and love as they have about the beauty of plants: Fiona H, Wendy, Jenny, Karen, Rhi, Andrea, Claire, Magali, Annie, Fiona, and Lizzie. Thank you for tucking me under your wings and imparting your aromatic wisdom. I would not be the clinical aromatherapist I am today without your enduring teachings, life philosophies, and unconditional love.

Thank you to my family for always calling me home. To Mummy and Daddy for instilling in me my work ethic and always supporting my next move, no matter how crazy or how many thousands of miles away. To my sisters, Donna, Kerrie, and Shannon, for giving me wings while always being my forever home. Your unwavering faith and confidence in me are what drives me forward. To the babies for reminding me that life is to be grasped with both hands and laughter is the only medicine that matters. To Bernardo for your inspiration, support, and love.

To Shannon, the conviction of your belief has been fundamental to me being the woman I am today. I couldn't do what I do without you.

To Alisa, that very wise woman who taught me the power of a village, I miss you dearly every day.